Practical Astronomy

Springer
London
Berlin
Heidelberg
New York
Barcelona
Hong Kong
Milan
Paris
Santa Clara
Singapore
Tokyo

Other titles in this series

The Observational Amateur Astronomer
Patrick Moore (Ed.)

The Modern Amateur Astronomer
Patrick Moore (Ed.)

Telescopes and Techniques
C.R. Kitchin

Small Astronomical Observatories
Patrick Moore (Ed.)

The Art and Science of CCD Astronomy
David Ratledge (Ed.)

The Observer's Year
Patrick Moore

Seeing Stars
Chris Kitchin and Robert W. Forrest

Photo-guide to the Constellations
Chris Kitchin

The Sun in Eclipse
Michael Maunder and Patrick Moore

Software and Data for Practical Astronomers
David Ratledge

Amateur Telescope Making
Stephen F. Tonkin

Observing Meteors, Comets, Supernovae and other
Transient Phenomena
Neil Bone

Astronomical Equipment for Amateurs
Martin Mobberley

Transit
When Planets
Cross the Sun

Michael Maunder and
Patrick Moore

With 35 Figures

Springer

ISBN 1-85233-621-8 Springer-Verlag London Berlin Heidelberg

British Library Cataloguing in Publication Data
Maunder, Michael
    Transit : when planets cross the sun. – (Practical
    astronomy)
    1. Transits
    I. Title   II. Moore, Patrick, 1923–
    523.9′9
ISBN 1852336218

Library of Congress Cataloging-in-Publication Data
Maunder, Michael J. de F.
    Transit : when planets cross the sun / Michael Maunder, [Patrick
Moore].
        p.   cm. – (Practical astronomy)
    Includes index.
    ISBN 1-85233-621-8 (alk. paper)
    1. Transits. I. Moore, Patrick. II. Title.   III. Series
QB175.M453   1999   99–31314
523.9′9–dc21   CIP

Typeset by EXPO Holdings, Malaysia
Printed and bound at the University Press, Cambridge
58/3830–543210   Printed on acid-free paper   SPIN 10677671

AAV– 6280

Contents

v

Introduction

There are some astronomical phenomena which are excessively rare. For example, the last time that the planet Venus passed across the face of the Sun, in transit, was in 1882; it will not happen again until 2004, so that there can now be no living person who can remember seeing a transit of Venus.

The only planets which are closer to the Sun than we are, and can therefore be seen in transit, are Mercury and Venus. During transit the planet appears as a black disk against the Sun's face; in the case of Venus, very striking as seen with the naked eye. In previous times, transits of Venus were regarded as very important astronomically, because they allowed astronomers to measure one very important quantity: the length of the astronomical unit, or distance between the Earth and Sun. The method was indirect, and was first proposed by Edmond Halley, of Halley's Comet fame. This is why the transits of the 17th, 18th and 19th centuries were so carefully observed.

Transit stories are legion!

There was the first observation of a Venus transit, made in 1639 by the brilliant but tragic young amateur Jeremiah Horrocks, when he could escape his clerical duties. There was the remarkable voyage of Monsieur Guillaume Legentil, who set out for India to observe the 1761 transit; arrived too late; stayed in India for eight years to observe the 1769 transit; was clouded out; suffered two shipwrecks on the way home, and on returning to France found that he had been presumed dead, so that his heirs were preparing to distribute his property.

There was the discovery of the atmosphere of Venus, at the 1761 transit, by the first great Russian astronomer, Mikhail Lomonosov. And, of course, there was the voyage of Captain Cook in 1769. He did not set out to discover Australia; he was sent by the Admiralty to go to Tahiti and make measurements of the transit of Venus.

There were the many problems of the 1874 and 1882 transits, mainly because of the dreaded "Black Drop" which ruined the accuracy of the observations....... Quite apart from Mercury and Venus, there was the curious story of the "ghost planet" Vulcan, predicted by one of the world's leading mathematicians, and allegedly seen in transit across the Sun by a French carpenter-cum-astronomer who used to record his observations on planks of wood, planing them off when he had no further use for them.

We now look forward to the 2004 transit of Venus, which should be well seen; before that we have a transit of Mercury in 1999. After 2012, over a century will pass before Venus again crosses the face of the Sun.

The book was of course planned and written jointly, though in fact one author (PM) was mainly responsible for the first 80 pages and the other author (MM) for the rest.

We are most grateful for the help given to us by the Publishers, particularly by John Watson.

June 1999

Part 1

Transits down the Ages

Chapter 1
The Moving Planets

For a few hours on 8 June 2004, astronomers all over the world will be very much on the alert. They will be watching the planet Venus silhouetted against the brilliant disk of the Sun – something which has not happened since 1882. Nobody living today can remember a transit of Venus. That of 2004 will be followed by another in 2012; after that astronomers must wait patiently until December 2117.

In bygone days these transits of Venus were regarded as being of special importance, because they could be used to measure the length of the astronomical unit – that is to say the distance between the Earth and the Sun, which, in round numbers, is now known to be 93 000 000 miles.* Admittedly this is no longer true, and much better and more accurate methods are now available, but transits are still fascinating to watch. Mercury, the other inner planet, can also pass in transit, and will do so next on 15 November 1999.

First, let us set the scene and say something about the make-up of the great universe in which we live.

The Sun is a star, and the stars are suns. Many of those tiny, twinkling points which can be seen on any clear, dark night are far larger, hotter and more luminous than our own Sun, but of course they are much further away. Light, travelling at about 186 000 miles per second, takes more than four years to reach us from even the nearest star, but to leap from the Sun

---

* In this book we have used Imperial units, so that everyone can understand what is meant. If you want to convert miles into kilometres, multiply by 1.609.

to the Earth takes a ray of light only eight and a half minutes.

The planets are different; they are the principal members of the Sun's family or Solar System, and are relatively close.

Mercury and Venus are closer to the Sun than we are, and are termed the inferior planets; the rest – the superior planets – are further out, and so take longer to complete one orbit. In general their paths are not very different from circles, though Pluto, the outermost member of the family, is an exception. The main details of the planetary system are best summarised in a table (see Table 1.1).

Some of the planets have satellites; we have only one – our familiar Moon – but Saturn has as many as eighteen. There can be no doubt that the Earth and the Moon have been associated since the Solar System was formed, around four and a half thousand million years ago, and it may be best to regard the Earth-Moon system as a double planet rather than as a planet and a satellite.

Like the Moon, the planets have no light of their own; they shine only by reflecting the rays of the Sun, so that if some malevolent giant suddenly snatched the Sun out of the sky, the planets would vanish too. (Let us assure you that this is not likely to happen.) Of course the Sun can light up only half of a globe at any one time, so that if a planet turns its dark of night hemisphere toward us we cannot see it at all. This happens with the Moon and with the two inferior planets, Mercury and Venus, but of course not with the rest. It is therefore obvious why Venus, seen against the face of the Sun, appears jet-black. So does Mercury, though Mercury is so much smaller than Venus and so much further away from Earth, that during transit it is well beyond naked-eye visibility.

One of the fundamental traffic laws of the Solar System is that the orbital speed falls off with increasing distance from the Sun, so that Mercury races along at almost 30 miles per second while the Earth is content with a modest 18.5 miles per second. Moreover, a planet moves at its quickest when closest to the Sun (perihelion) and slowest when it is at its furthest (aphelion). "Magnitude" refers to a planet's apparent brightness; the lower the magnitude, the brighter the object. Venus, at –4.4, can be seen even when the Sun is above the horizon, but Uranus is barely visible without optical aid, and to see Pluto at all you need a telescope of fair size.

The Solar System is divided into two very definite parts. First come the four "terrestrial" planets, Mercury, Venus, the Earth and Mars, which are solid and rocky, and relatively dense. Beyond Mars comes a wide gap, in which move many thousands of dwarf worlds known as asteroids or minor planets; only one (Ceres) is as much as 400 miles in diameter, and only one (Vesta) is ever visible with the naked eye. Further out come the four giants, Jupiter, Saturn, Uranus and Neptune, which have gaseous surfaces and all of which have extensive satellite families. Finally there is Pluto, a strange little world smaller than the Moon, whose orbit can bring it within that of Neptune, though for most of its 248-year period it is much further out. In recent years many asteroid-sized bodies have been found in these remote parts of the Solar System, and it may be that Pluto is merely the largest of an outer swarm. Its status remains uncertain; it simply does not seem to fit into the general pattern.

Drawing a plan of the Solar System is not too easy, because of the scale involved. Neptune, the outermost of the main planets, is almost 80 times further out than Mercury. On the other hand, drawing the plan on a flat piece of paper is much less of a problem, because most of the planets move in roughly the same plane. Relative to the Earth's orbit, the inclinations of the paths of the other planets are less than three degrees, apart from Mercury and Venus (not counting Pluto, which probably does not merit true planetary status). Yet it is the slight but definite inclinations of the orbits of Mercury and Venus which are so important when we come to consider the ways in which transits occur.

Let us first deal with the two inferior planets. We can concentrate on Venus, but Mercury behaves in essentially the same way.

In the diagram, the orbits of Venus and Earth are shown (not to scale). Begin with Venus in position 1. Its dark or night side is turned toward us, and we cannot see it, unless it is directly in front of the Sun and appears in transit. Venus is "new"; this is termed inferior conjunction. Venus then moves on in its orbit, and we see it as a thin crescent in the morning sky. By the time it has reached position 2, half of its daylit side is turned toward us, and so Venus appears as a half disk; this is known as dichotomy. Between positions 2 and 3 Venus is gibbous – that is to say, between half and full. At position 3 we face the whole of the sunlit hemisphere, and Venus is full, but as it is then on the

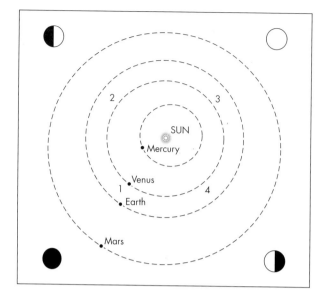

Fig. 1.1. The appearance of Venus (and Mercury) at different points of the orbit.

far side of the Sun it is to all intents and purposes unobservable (superior conjunction). It then emerges into the evening sky, at first gibbous, then half (position 4) and next a narrowing crescent until it returns to new (position 1). It takes Venus nearly 225 days to complete one journey round the Sun, but the interval between one inferior conjunction and the next – the synodic period – is almost 584 days.

The greatest angular distance between Venus and the Sun is nearly 50 degrees, so that Venus can therefore remain above the horizon for several hours after sunset or rise several hours before the Sun does so. Apart from the Sun and the Moon, it is much the brightest object in the entire sky; at its best it looks rather like a small lamp, and can cast perceptible shadows. The apparent diameter ranges between over 65 seconds of arc at maximum, down to only 9.5 seconds of arc at minimum. Of course Venus is closest to us, and has its greatest apparent diameter, when at inferior conjunction and is therefore out of view except during a transit. As the phase increases, so the apparent diameter shrinks. In some ways Venus is an infuriating object to observe, particularly since its surface is permanently hidden by its dense, cloud-laden atmosphere.

En passant, this atmosphere, which is much more extensive than that of the Earth, has an effect upon the phase. When Venus is an evening object, and therefore waning, dichotomy is always rather earlier than theory predicts; during morning elongations, with a waxing Venus, dichotomy is late. This was first noticed by Johann Schröter, in the 1790s, and is now generally known as the Schröter effect.[*] It is due solely to the planet's atmosphere, and there is no suggestion that the theoretical calculations are wrong. A list of elongations and conjunctions of Venus for the period 1999–2005 is given in the Appendix (page 161).

Venus can approach us to within 25 000 000 miles; Mercury, a much smaller globe, can never come closer than 49 000 000 miles. Moreover, Mercury's angular distance from the Sun can never be more than 29 degrees. The synodic period is 115. 9 days, so Mercury is a quick mover; it was appropriately named in honour of the messenger of the Olympian gods. The main problem about observing it is that it always stays in the same part of the sky as the Sun, so that with the naked eye it can be glimpsed briefly only when very low in the west after sunset or else very low in the east before dawn. It is never obtrusive, and there are many people who have never seen it at all; in fact, you are almost certain to overlook it unless you make a deliberate search to it. Mercury has virtually no atmosphere, so that there is no Schröter effect.

Before the Space Age, the best maps of Mercury were made by astronomers who observed the planet when both it and the Sun were high in the sky; this, naturally, involved using telescopes with very accurate setting circles. Even then, the maps proved to be very rough, and we had to wait for the results from an unmanned space-craft, Mariner 10, which made three active passes of Mercury in the 1970s and sent back our first detailed information from close range.

The superior planets behave quite differently. In the next diagram the orbits of the Earth and Mars are shown, again not to scale. With the Earth at E1 and Mars at M1, it is clear that the two planets and the Sun are lined up; as seen from Earth, Mars is exactly opposite to the Sun in the sky, and is said to be at opposition, while to an observer on Mars the Earth would be at inferior conjunction. With perfect alignment, our Martian watcher would see a transit of the Earth. A year later the Earth has completed one circuit of the Sun, but Mars, moving more slowly in a larger orbit, has not had

---

[*] I gave it this nickname, quite casually, during a television *Sky at Night* programme more than thirty years ago. To my surprise, everyone now seems to use it! – PM

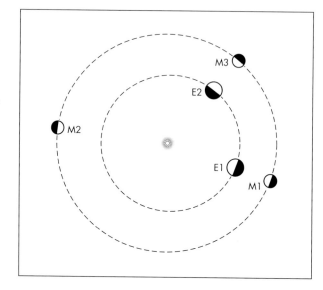

Fig. 1.2. Why Mars cannot see the Earth transit the Sun every year.

time to do so; it has only reached M2. The Earth has to catch Mars up, so to speak, and this happens at the next opposition, with the Earth at E2 and Mars at M3. The synodic period of Mars – that is to say, the interval between successive oppositions – is 780 days; thus oppositions fall in 1997 and 1999, but not in year 1996 and 1998. Note also that although Mars can appear gibbous when at its maximum elongation from the Sun, it is never less than 86% illuminated as seen from Earth.

The outer planets, Jupiter and the rest, have shorter synodic periods, because they are more easily caught up by the Earth, and their phases are inappreciable.

Mercury, with a diameter of little more than 3000 miles, is not a great deal larger than the Moon, though admittedly it is much more massive. It has a relatively weak gravitational pull, and this means that it has not been able to retain much in the way of atmosphere; when we call Mercury "an airless world" we are not far wrong. There is an iron-rich core, as with the Earth, and there is a definite magnetic field. The axial rotation is very slow. Mercury takes over 58 days to spin once round, so that one Mercurian "day" amounts to two-thirds of a Mercurian "year", and the calendar there is judged very strange by terrestrial standards. Once the Sun rises over any particular location, it will not set again for a period equal to 88 Earth days.

Mariner 10 showed us that Mercury is a world of craters, mountains, valleys and plains. There is one huge ringed structure, the Caloris Basin, which is over

800 miles in diameter, and is bounded by a ring of mountain blocks which rise to as much as a mile and a quarter above the adjacent surface. The maximum temperature at noon, in the hottest part of the planet, is over 700 degrees Fahrenheit, though the nights are bitterly cold. Mercury is a hostile place, and the chances of any manned expeditions there in the foreseeable future seem to be nil.

Venus is as different from Mercury as it could possibly be. In size and mass it is almost a twin of the Earth (if you represented them both by snooker balls, you could quite easily use them for play), but there the resemblance ends. Venus' thick atmosphere, with a ground pressure at least 90 times that of the Earth's air at sea level, is made up chiefly of the heavy, unbreathable gas carbon dioxide, while the clouds are rich in sulphuric acid, and the surface temperature is not far short of 1000 degrees Fahrenheit. The rotation period is slow, and indeed it takes 243 Earth days for Venus to spin round, so that the "day" there is technically longer that the "year" of 224.7 Earth days. To make the situation even more curious, Venus rotates from east to west, not from west to east. This means that if it were possible to have a clear view from the surface, the Sun would rise in the west and set in the east 118 Earth days later. Why Venus spins in this peculiar way is not known. The mean density of the globe is rather less that that of the Earth, and no magnetic field has been detected.

Before the Space Age it was widely believed that Venus might be a welcoming world, perhaps with broad oceans and even life. We now know that nothing could be further from the truth. Unmanned vehicles have been sent there and put into closed paths round the planet, so that the surface could be mapped by radar; there have also been actual controlled landings, and pictures have been sent back, though no probe can survive in an active condition there for more than a few hours before being permanently silenced. The carbon dioxide atmosphere acts in the manner of a greenhouse, and blankets in the Sun's heat, which is why the surface temperature is so high.

Craters are everywhere, with extensive lava-flows; there are mountains, valleys, and volcanoes which are almost certainly active. A huge rolling plain covers over 60% of the surface, and there are two major upland areas – the highest peaks rise to five miles above the surrounding areas. Though Venus is beautiful when seen with the naked eye, and was named after the

Goddess of Love, conditions there are much more akin to the conventional idea of hell. Life of any kind seems to be completely out of the question.

We are fairly sure that the planets were formed, around 4.5 thousand million years ago, from a cloud of dust and gas surrounding the youthful Sun. In those early times the Sun was not so luminous as it is now, and Venus and Earth may well have started to evolve along similar lines, with extensive oceans. But then the Sun became more powerful, with devastating effects. Earth, at a mean distance of 93 000 000 miles, was more or less out of harm's way; Venus, at only 67 000 000 miles, was not so lucky. There was a runaway greenhouse effect; the oceans boiled away, the carbonates were driven out of the rocks, and in a comparatively short time by cosmical standards Venus changed from a potentially life-bearing world into the scorching hot desert of the present time. It is rather sobering to reflect that if the Earth had been a mere 20 million miles closer to the Sun it would have suffered the same fate, and you would not now be reading this book.

Mercury and Venus are unattended, but Earth, of course, has one satellite: the Moon, which moves at a mean distance of only 239 000 miles, and completes one orbit in 27.3 days, during which time it goes through the complete cycle of phases from new to full. Even with the naked eye it is easy to see the broad plains which are still mis-called "seas" even though there has never been any water in them, while binoculars will reveal the mountain and craters, and with a telescope the views are magnificent. There is practically no atmosphere, and no overall magnetic field.

The orbital inclination, referred to the orbit of the Earth, is 5 degrees 9 minutes. But for this, the Moon would pass in front of the Sun every time it returned to "new", and we would have solar eclipses every month. In fact this does not happen; on most occasions the new moon passes unseen either above or below the Sun in the sky, and there is no eclipse. It is sheer chance that the Sun and Moon appear virtually the same size in the sky; the Sun's diameter is 400 times greater than that of the Moon, but the Sun is also 400 times further away.

A total solar eclipse – when the alignment is exact and the Moon completely hides the brilliant disk of the Sun – is surely the grandest sight in all nature; for a few minutes we can see the Sun's atmosphere, together with the masses of glowing gas once called Red Flames and now called prominences, as well as the glorious pearly corona, which extends outward from the Sun in

Table 1.1. The Planets

Planet	Distance from Sun, millions of miles			Orbital period	Synodic period (days)	Orbital eccentricity	Orbital inclination (degrees)
	Max	Mean	Min				
Mercury	43	36	29	88 days	115.9	0.206	7.0
Venus	67.7	67.2	66.7	224.7 days	583.9	0.007	3.4
Earth	94.5	93.0	91.4	365.3 days	–	0.017	0
Mars	154.8	151.7	128.5	687 days	779.9	0.09	1.9
Jupiter	507	484	460	11.9 years	398.9	0.048	1.3
Saturn	937	887	837	29.5 years	378.1	0.056	2.5
Uranus	1867	1783	1699	84.0 years	369.7	0.047	0.8
Neptune	2820	2795	2769	164.8 years	367.5	0.009	1.8
Pluto	4583	3666	2750	247.7 years	366.7	0.248	17.1

all directions. Since the Moon's shadow is only just long enough to reach the Earth, total eclipses are rare as seen from any particular location from England – the last occasion was in 1927, the next will be in August 1999, and after that we must wait until 2090 (though, of course, many total eclipses will be seen before then from other parts of the globe). This is no place to go into details about eclipses of the Sun, particularly as we have done so elsewhere,[*] but it is worth noting that the term is technically wrong, because a solar eclipse is really an occultation of the Sun by the Moon – or, if you like, a transit of the Moon.

Table 1.2. Apparent sizes from Earth

	Object (minutes)	Angle (seconds)
Sun	32	35
Mercury	00	12.9
Venus	00	65.2
Mars	00	25.7
Jupiter	00	50.1
Saturn	00	20.9
Uranus	00	3.7
Neptune	00	2.2
Pluto	00	less than 0.1 in year 2000

These are the apparent sizes of the Sun and main lanets when at their closest to Earth.
For Pluto the figure will always be less than 0.1 arcsecond.

---

[*] *Sun in Eclipse*, by Michael Maunder and Patrick Moore; Springer-Verlag, 1998.

Neither need we spend much time here on the outer planets, but it is worth at least mentioning their satellites, some of which are large. Jupiter has four major satellites: Io, Europa, Ganymede and Callisto, which can often be seen in transit against the disk of the planet; a very modest telescope will suffice. Saturn has one large attendant, Titan, and several more of moderate size (Iapetus, Rhea, Dione, and Tethys). The satellites of Uranus and Neptune are much more difficult to see, and reasonably large telescopes are required to show them at all.

By now we have sent rockets out to all the planets apart from Pluto, and controlled landings have been made on Mars and Venus. Only the Moon has so far been reached by astronauts, but no doubt the first expeditions to Mars will set out at some time during the first part of the coming century. The direct exploration of this Solar System is well under way.

Chapter 2

Transits: Why and When?

Figure 2.1.
Circumstances for observing a transit of Venus (or Mercury) M is mid-transit for the theoretical best circumstances. These are for a point on the Earth's Equator and a bisecting transit across the solar equator. *IS*: Ingress at the Earth's South Pole. *IN*: Ingress at the Earth's North Pole. *ES*: Egress at the Earth's South Pole. *EN*: Egress at the Earth's North Pole.

We have seen that if we draw a plan of the Solar System on a flat piece of paper, we are not very far wrong; but the slight inclinations of the orbits of Mercury and Venus make all the difference insofar as transits are concerned. If their orbits were in the same plane as ours, the planet would transit at every inferior conjunction – that is to say every 116 days for Mercury and every 584 days for Venus. In fact this does not happen, and transits are comparatively rare.

Let us consider Venus first, where the inclination of the orbit relative to ours, is 3.4 degrees. We can visualise the orbits as resembling two hoops which are inclined to each other by this amount. The points where the two hoops cross are called the nodes. Half of Venus' orbit is "above" ours, and the other half "below" (technically one ought to say "north" and "south", because there is no fixed point of reference in space, and so no

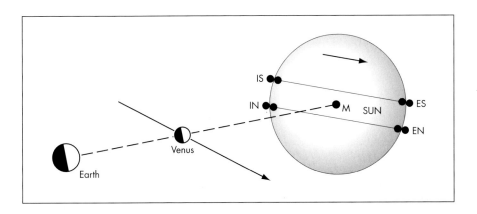

true "up" or "down"). Venus takes 224.7 Earth days to go once round the Sun. For a transit to occur, Venus must be at, or very near, a node at the time of inferior conjunction; otherwise it will pass either above or below the Sun in the sky, and will escape transit.

The plane of the orbit of Venus passes through that of the Earth twice a year: on 6 June and 7 December. This means that a transit can occur only on, or very near, one or other of those dates. There is a certain amount of leeway, but not much, because the orbit of Venus is very nearly a circle; the orbital eccentricity is only 0.007, whereas that of the Earth's orbit is 0.017.

It so happens that eight Earth years are very nearly equal to thirteen Venus years, so that after eight years the two planets and the Sun are in nearly the same relative positions (the vital word here is "nearly"). A transit may therefore be followed by another transit eight years later, as happened in 1874 and 1882. But in 1890, when the relative positions were again nearly the same, the alignment was no longer exact enough for a transit to occur, and observers had to wait impatiently for the next pair, due in 2004 and 2012.[*]

To sum up: at the present epoch transits of Venus occur in pairs, separated by over a century. At least a transit of Venus is a leisurely affair, and there is no need for frantic preparations as there is for a total eclipse of the Sun. For example, at the next transit, that of 8 June 2004, Venus begins to draw on to the Sun at 5h 15m GMT and does not finally leave it until 11h 28m, giving an observer more than six hours to avoid cloud-cover even or even to hire an aircraft.

Table 2.1 lists the transits of Venus between the years 1500 and 2200. First contact indicates the time when Venus first touches the Sun's limb, last contact gives the time when it leaves – reckoned from the view point of an observer at the Earth's centre; obviously observers at different locations will have somewhat different views, which is very important astronomically. Previously there had been transits in 1518 and 1526, but so far as we know they were not observed (neither, for that matter, was the transit of 1631, though it had

---

[*] A very rough calculation will show that the relationship is close. Take the length of Venus' orbital period as 224.701 days, and that of the Earth 365.256 days. $224.701 \times 13 = 2921.113$. $365.256 \times 8 = 2922.048$. This discrepancy amounts to less than one day.

Table 2.1. Transits of Venus between the years 1600 and 2200

Date	First contact	Mid-transit	Last contact	Duration
1631 Dec. 7	03.49	05.19	06.49	3h 0m
1639 Dec. 4	14.56	18.25	21.54	6h 58m
1761 June 6	02.01	05.19	08.36	6h 35m
1769 June 3/4	19.15	22.25	01.35	6h 20m
1874 Dec. 9	01.50	04.08	06.26	4h 36m
1882 Dec. 6	13.57	17.07	20.16	6h 13m
2004 June 8	05.15	08.21	11.28	6h 13m
2012 June 5/6	22.22	01.33	04.53	6h 31m
2117 Dec. 11	00.03	02.53	05.44	5h 41m
2125 Dec. 8	13.19	16.07	18.54	5h 35m

been predicted). All times are in GMT, and all dates follow the new-style calendar.

Then follow 11 June 2247, 9 June 2255, 12–13 December 2360 and 10 December 2368. Obviously, all transits after that of 2012 are of academic interest only so far as we are concerned.[*]

The situation with Mercury is much less straightforward, because Mercury moves much more quickly round the Sun than Venus does, and because its orbital eccentricity is much greater (0.206). Orbital plane crossings occur in May and November, and these are the only months when transits can occur. Near the May crossing, Mercury is not far from its aphelion point

Table 2.2. Transits of Mercury between the years 1600 and 1995

1605 Nov. 01	1707 May 05	1799 May 07	1894 Nov. 10
1615 May 03	1710 Nov. 06	1802 Nov. 09	1907 Nov. 14
1618 Nov. 04	1723 Nov. 09	1815 Nov. 12	1914 Nov. 07
1628 May 05	1736 Nov. 11	1822 Nov. 05	1924 May 08
1631 Nov. 07	1740 May 02	1832 May 05	1927 Nov. 10
1644 Nov. 09	1743 Nov. 05	1835 Nov. 07	1937 May 11
1651 Nov. 03	1753 May 06	1845 May 08	1940 Nov. 11
1661 May 03	1756 Nov. 07	1848 Nov. 09	1953 Nov. 14
1664 Nov. 04	1769 Nov. 09	1861 Nov. 12	1957 May 06
1674 May 07	1776 Nov. 02	1868 Nov. 05	1960 Nov. 07
1677 Nov. 07	1782 Nov. 12	1878 May 06	1970 May 09
1690 Nov. 10	1786 May 04	1881 Nov. 08	1973 Nov. 10
1697 Nov. 03	1789 Nov. 05	1891 May 10	1986 Nov. 13
			1993 Nov. 06

---

[*] These are the values given by Jean Meeus, the Belgian astronomer, whose tables are absolutely invaluable.

Table 2.3. Transits of Mercury between the years 1999 and 2050

Date	First contact	Mid-transit	Last contact	Duration
1999 Nov. 15	21.16	21.42	22.08	0h 52m
2003 May 7	05.15	07.54	10.33	5h 17m
2006 Nov. 8/9	19.13	21.42	00.11	4h 58m
2016 May 9	11.14	14.59	18.44	7h 30m
2019 Nov. 11	12.37	15.21	18.06	5h 31m
2032 Nov. 13	06.42	08.55	11.09	4h 51m
2039 Nov. 7	07.20	08.48	10.17	2h 57m
2049 May 7	11.06	14.26	17.46	6h 40m

(furthest distance from the Sun), whereas in November it is near its perihelion, or closest distance to the Sun. This means that November transits are more frequent than those of May in the ratio of 9 to 4. The durations vary; a transit can last for as little as a few minutes (as in 1937) or for as long as several hours. There seems little point in giving full details about transits which took place long ago, but Table 2.2 shows the dates between 1600 and 1995.

Then follow November 11 2052, May 5 2062, November 11 2065 and November 14 2078.

It cannot honestly be said that there is anything spectacular about a transit of Venus, and even less so for Mercury. Yet in past years transits were regarded as being of special importance, due initially to a proposal made by one of the greatest and most charismatic of all British astronomers: Edmond Halley. To him we now turn.

Chapter 3
Edmond Halley and the Distance of the Sun

Of all the personalities in the history of astronomy, Edmond Halley is surely one of the most charismatic. But for the fact that he was contemporary with Isaac Newton, he would be remembered as the greatest astronomer of the 17th and 18th centuries. Mathematically, of course, he was not the equal of Newton – who was? – but in every other way Halley was superior. We remember him best in connection with the comet which bears his name, and which last visited us in 1986, but this was only one of Halley's achievements, and a very minor one at that. He was very much in the mould of what we would today call an all-rounder; his interests ranged from pure science to surveying, military fortifications and even undersea exploration – he went so far as to build, and test, a primitive sort of submarine. It was Halley who persuaded Newton to write the immortal *Principia*, and he even paid for its production. He became Astronomer Royal in 1720, in succession to John Flamsteed, and remained in the post until his death in 1742.

Unlike some of his contemporaries (notably Flamsteed and Newton) he was gregarious and jovial, with a very strong sense of humour. There is a story that when the Czar of Russia, Peter the Great, visited England to learn about shipbuilding, he struck up a friendship with Halley, and a far from teetotal evening ended when the Czar climbed into a wheelbarrow and Halley pushed him through a hedge. The story may be apocryphal, but it is certainly true that the records of Greenwich Observatory include a receipt for two damaged wheelbarrows... The puritanical Flamsteed complained that Halley "swore and drank brandy like a sea-captain". No doubt he did, and, typically his last action on earth was to call for a glass of wine – and drink it.

Halley was born in 1656, and was educated at St. Paul's School, becoming school captain in 1671. Two years later he entered Queens College, Oxford, with the promise of a brilliant career ahead, but before completing his degree he went to the island of St. Helena in order to draw up the first good catalogue of the southern stars, which can never be seen from the latitudes of Europe. Despite bad weather and an unco-operative island governor, he carried out the project so well that he was nicknamed "the southern Tycho", and on his return he was at once awarded an honorary degree. But from our present point of view, the most important fact of his trip to St. Helena is that while he was there, on 7 November 1677, he observed a transit of Mercury. He timed it very accurately, and this led him on to suggest that transits of the inferior planets might be used to measure the length of the astronomical unit, or Earth–Sun distance. A vague suggestion of the same kind had been made earlier by James Gregory, but it was Halley who worked it out in detail.

In 1716 he published a paper in the Proceedings of the Royal Society which is worth quoting here:

> Scarce any problem will appear more hard or difficult than that of determining the distance of the Sun from the Earth; but even this, when we are made acquainted with some exact observations taken at places fixed upon and chosen beforehand, will, without much labour, be effected. And this is what I am now desirous to lay before this illustrious Society (which I foretell will continue for ages), that I may explain beforehand to young astronomers, who may perhaps live to observe these things, a method by which the immense distance of the Sun may be truly obtained to within a five-hundredth part of what it really is.
>
> It is well known that the distance of the Sun from the Earth as by different astronomers supposed different; according to what was judged most probable from the best conjecture that each could form … But the planets Venus and Mercury having, by the assistance of the telescope, been seen in the disk of the Sun, deprived

of their borrowed brightness, it is at length found that the apparent diameters of the planets are much less than they were formerly supposed...

About forty years ago, when I was in the island of St. Helena, observing the stars about the south pole, I had an opportunity of observing, with the greatest diligence, Mercury passing over the disk of the Sun; and (which succeeded better than I could have hoped for) I observed, with the greatest degree of accuracy, by means of a telescope 24 feet long, the very moment when Mercury, entering upon the Sun, seemed to touch its limb within, and also the moment when going off if struck the limb of the Sun's disk, forming the angle of interior contact; whence I found the interval of time, during which then appeared within the Sun's disk, even without an error of one second of time. For the lucid line intercepted between the dark limb of the planet and the bright limb of the Sun, although exceedingly fine, is seen by the eye, and the little dent made on the Sun's limb, by Mercury's entering the disk, appears to vanish in a moment, and also that made by Mercury leaving the disk seems to begin in an instant. When I perceived this it immediately came into my mind that the Sun's parallax might be accurately determined by such kinds of observations as these, provided Mercury were but nearer to the Earth, and had a greater parallax from the Sun; but the difference of these parallaxes is so little as always to be less than the solar parallax which we seek, and therefore Mercury, though frequently to be seen on the Sun, is not to be looked upon as fit for our purpose.

There remains, the transit of Venus over the Sun's disk; whose parallax, being almost as great as the solar parallax, will cause very sensible differences between the times when Venus will be seen to be passing over the Sun at different parts of the Earth. And from these differences, if they be observed as they ought, the Sun's parallax may be determined even to a small part of a second. Nor do we require any other instruments for this purpose than common telescopes and clocks, only good of their kind; and in the observers nothing more is needful than fidelity, diligence, and a moderate skill in astronomy. For there is no need that the latitude of the placed should be scrupulously observed nor that the hours themselves should be accurately determined with the respect to the meridian; it is sufficient that the clocks be regulated according to the motion of the heavens, if the times be well reckoned from the total ingress of Venus into the Sun's disk to the beginning of her egress from it; that is, when the dark globe of Venus first begins to touch the bright limb of the Sun within; which moments I know, by my own experience, may be observed within a second of time.

But on account of the very strict laws by which the motions of the planets are regulated, Venus is seldom seen within the Sun's disk and during the course of 120 years it could not be seen once – namely, from the year 1639 (when this most pleasing sight happened to that excellent youth, Horrocks, our countrymen, and to him only since the Creation)[*], to the year 1761, in which year, according to the theories which we have hitherto found agreeable to the celestial motions, Venus will again pass over the Sun on May 26 (6 June, new style) in the morning.

Halley goes on to explain that the duration of the transit will be different as seen from different parts of the Earth, so that the method involved making observations from sites as widely separated in latitude as possible – say, north Norway and the East Indies. From these durations, it will be mathematically possible to obtain the solar parallax. He continues: "I recommend it therefore again and again to those curious astronomers who (when I am dead) will have an opportunity of observing these things."

Halley's reasoning is straightforward enough. The main problem lay in making the observations, because it involved sending expeditions to very widely separated localities – and that was much more difficult in Halley's time than it is now. It also meant accurate timing of both ingress and egress, and weather conditions could never be relied upon. A modification was later proposed by French astronomer Joseph Nicolas Delisle, who actually visited England to discuss the

---

[*] In fact the 1639 transit was also seen by William Crabtree. See page 30 and 36.

whole problem with Halley. Delisle pointed out that an observer who sees the planet take a longer path across the Sun must see ingress earlier than an observer who sees the planet take a shorter path. (After all, a longer path must mean an earlier ingress and a later egress.) Therefore it is only necessary to time the exact moment of either ingress and egress. Then, from the exact knowledge of the position of the observing site, the time difference between two points of observation would be proportional to the time by which the contact for the longer path precedes the contact for the shorter path, and hence proportional to the distance between them. The size of the Sun, and hence its distance, follows.

Observationally, Delisle's method was easier than Halley's; only one timing was needed – either ingress or egress – and sites could be chosen where either of these, but not both could be seen. On the debit side, it required a precise knowledge of the longitude of the observing site, and this sort of information was often lacking in the 18th century.

Halley, as we have seen, realised that Mercury was not a suitable target. Initially Delisle thought otherwise, but later he too came round to Halley's view, and great hopes were centred upon the forthcoming transit of Venus in 1761. But before coming on to the observations of that year, let us dispose of the less important and less spectacular transits of Mercury.

Chapter 4
Transits of Mercury

A 12th-century Moroccan astronomer, Abu Ishaq Al-Bitruji Al-Ishbili – better remembered today as Alpetragius – made the comment that since he had never seen Mercury pass across the face of the Sun, the planet must be transparent. What he did not know was that when in transit, Mercury is too small to be seen with the naked eye – and there were no telescopes in Alpetragius' day.

So it was not until 1631 that the first transit of Mercury was actually observed.

It had been predicted by Kepler, on the basis of his new Rudolphine Tables of planetary motions. Telescopes had come on the scene around 1610, but the next few transits of Mercury, in 1615, 1618 and 1628, had passed unseen because nobody knew that they were due. Things were different in 1631, and the French astronomer Pierre Gassendi made up his mind to catch Mercury in the act. He did so, and wrote:

> The crafty god had sought to deceive astronomers by passing over the Sun a little smaller than was expected, and had drawn a veil of dark clouds over the earth, in order to make his escape more effectual. But Apollo, acquainted with his knavish tricks from infancy, would not allow him to pass altogether unnoticed. To be brief, I have been more fortunate than those hunters after Mercury who have sought the cunning god in the Sun; I found him out, and saw him where no one else had hitherto seen him.

There are no reports of the transit of 1644, but that of 1651 was observed by Jeremiah Shakerley, about whom our information is decidedly scanty. He was certainly born in Halifax in November 1626, and although he did not attend University it is clear that he became a competent mathematician; at an early stage he recognised the importance of the work of Jeremiah Horrocks, whom we will meet in the next chapter of this book. Shakerley corresponded with many astronomers of the time, including another rather shadowy character, Richard Towneley, who is said to have made improvements to the astronomical micrometer invented by William Gascoigne; it may even be that for while Shakerley lived with Towneley's family in order to improve his knowledge of science.

Around 1650 Shakerley went to India, for the specific purpose of observing the 1651 transit of Mercury. He seems to have been successful, but no details have survived, and nothing more is heard of Shakerley after 1655, when he was still in India. It is believed that he died around 1670.

The next transit was that of 1661, and was more efficiently observed. In particular it was seen by the Dutch scientist Christiaan Huygens, who was possibly the best observer of the time (he discovered Titan, the brightest satellite of Saturn, as well as the true nature of Saturn's rings, though he is best remembered as the inventor of the pendulum clock). The transit was also seen by Hevelius of Danzig (the modern city of Gdaňsk), who had produced one of the earliest telescopic maps of the Moon, and had also drawn up an elaborate star catalogue. Hevelius commented that the planet looked much smaller than he had expected, and indeed his estimate agrees quite well with modern values; when at its closest to us, the apparent diameter of Mercury is almost 3 seconds of arc.

No records exist of the transits of 1664 or 1674, but then we come to 1677 – and Edmond Halley.

As we have noted, Halley had gone to the island of St. Helena in order to draw up the first good catalogue of the southern stars, which can never be seen from Europe. He had taken with him a telescope of focal length 24 feet, and he carried out his task well, despite unfavourable weather and an officious island governor. It was fortuitous that a transit of Mercury occurred during his stay. He made careful measurements, and wrote that he had "very accurately obtained the very moment in which Mercury, entering the Sun's limb, seemed to touch it internally … Hence I discovered the precise quantity of time … and that without an error of a single second." It was then that he realised that transits could well be used to measure the length of the astronomical unit.

However, he admitted that the disk of Mercury was too small to be measured precisely enough to yield the results he wanted, and he felt that Venus would be much more suitable, partly because it was closer and partly because it looked so much larger.

Unfortunately the next transit of Venus was not due until 1761, by which time Halley would have reached the advanced age of 108, so that all he could do was to lay down the procedure to be followed by his successors. Rather surprisingly, not everyone agreed that Venus would be more suitable than Mercury, and in particular William Whiston, who was a capable mathematician even though his religious views were somewhat peculiar, was firm in maintaining that Mercury should be the prime target. Whiston's comments were taken more seriously in France than in England, and French observers took care to make observations of the transits of 1723 and 1753. What they found showed that Halley had been right and Whiston wrong. Henceforth all the emphasis was on transits of Venus, and those of Mercury were more or less disregarded from this point of view.

All the same, most transits since then have been observed, and various investigations have been carried out, mainly concentrated upon three line of research:

1. The search for a possible satellite, accompanying Mercury across the Sun's face.
2. The effects of any Mercurian atmosphere, which would make the outline of the planet appear hazy instead of sharp (as does in fact happen with Venus).
3. Any phenomena on the disk of Mercury itself.

The first of these can be dismissed at once. We now know that Mercury is a solitary traveller in space, and moreover any satellite would have been so tiny that detecting it would have been well nigh impossible. But the other two investigations deserve more serious consideration.

Until fairly modern times it was thought that Mercury might have an atmosphere – much thinner than ours, but still appreciable. In 1934 E.M. Antoniadi, the leading planetary observer of the time, maintained that the atmosphere was thick enough to support clouds, and in this case it could easily have shown up during transits. As early as the transit of 1736 the French observer de Plantade reported a luminous ring round the black disk, and the same appearance was seen in 1799 by several observers, one of whom was Johann Hieronymus Schröter, who is regarded as the real "father of selenography" and whose drawings of the Moon and planets were far better than any previously made. In 1832 the same ring was reported, this time with a violet tinge. Another observation of it was made in 1868 by William Huggins, one of the pioneers of stellar spectroscopy, who saw an aureole of light round the disk as well as a luminous point on Mercury itself. According to Huggins, the breadth of the aureole was as much as half that of Mercury's angular diameter. There was no fading off at the margin, and both aureole and patch remained visible throughout the transit.

This is interesting, but we now know that the Mercurian atmosphere cannot be responsible. It is far more tenuous than Antoniadi believed, and it was not detected at all until 1974, when the unmanned spacecraft Mariner ten flew past the planet at a range of 470 miles and sent back detailed pictures as well as a vast amount of miscellaneous information. The atmosphere turned out to be remarkably thin, with an estimated ground density of 1/10 000 000 000 of a bar; this corresponds to what we normally call an excellent laboratory vacuum. In fact the atmosphere is absolutely negligible, and could not possibly produce any observable effects.

What else could produce an aureole, it is difficult to say, and it does seem that we are dealing with phenomena which are either due to diffraction, or else to observational or instrumental error.

When Venus draws on to the Sun at the start of a transit it seems to draw a strip of darkness after it, and when this disappears the transit has already begun; this is the infamous "Black Drop", to be discussed below. No doubt the atmosphere of Venus is mainly responsible. Nothing so definite has been seen with airless Mercury, and occasional apparent deformations of the disk must be put down to tremors in our own atmosphere. Antoniadi saw a kind of "Black Drop" effect during the transit of 1927, when he was using the 33-inch refractor at the observatory of Meudon, outside Paris – one of the best telescopes of its kind in the world – but he wrote that "the effects of diffraction and irradiation are, in general, completely drowned in that of atmospheric agitation, which enlarges the solar disk, making Mercury look smaller, and giving the limb a breadth great enough to cause this phenomenon, which is further exaggerated by the 'jumpiness' due to heat".

The first report of a bright spot on the Mercurian disk during a transit dates back to 1697, when it was noted by the German astronomer Johann von Würzelbau. During the 1799 transit Schröter and his assistant, Karl Harding, saw greyish spots on the disk, and similar observations followed at later transits. Huggins, known to be a particularly good observer, drew a bright speck on the planet, well away from the centre of the disk, during the transit of 1868, and so did a leading Australian astronomer, John Tebbutt, during the transit of 1881.

Yet all reports of this kind – and there were quite a number of them – are old, and this must surely be significant. Similar patches have not been seen more recently, despite the tremendous improvements in telescopic equipment and in techniques. After all, Mercury appears as a tiny disk against a brilliant background, so that observations are difficult; moreover the seeing conditions are generally unsteady. All in all, there can be no serious doubt that the reported patches are due to mere optical effects, and have nothing to do with Mercury itself.

Transits of Mercury may not be scientifically important, but they are always worth watching. The next chance will come on 15 November 1999. Unfortunately this transit will not be visible from Britain, but there will be a good opportunity for British observers in May 2003 – so let us hope for clear skies.

Chapter 5
Venus: The Transit of 1639

When Mercury crosses the Sun it is too small to be seen with the naked eye, and so there was no chance of observing a transit before telescopes were invented. Not so with Venus, which is so much larger and closer that there is no difficulty in seeing it against the Sun. It may be that some ancient-world or mediæval observer did in fact see a transit; there is a vague report of a sighting by an Arab star-gazer in 639, but it is very indefinite, and what he saw (if anything) could so easily have been a sunspot. Therefore, the first reliable records go back no further than the 17th century.

As we have noted, Johannes Kepler had predicted that both Mercury and Venus would transit in 1631, Mercury on 7 November and Venus on 6 December.* The French astronomer Pierre Gassendi set out to observe both. He was successful with Mercury, and so there seemed every reason to hope that he would be equally successful with Venus, and he left nothing to chance. Fearful that Kepler's timing might be in error, he began watching the Sun on 5 December, between breaks in the clouds, and went on throughout 6 and 7 December. To his surprise and disappointment, he saw nothing. What he did not know, and could not know, was that the transit took place during the northern night of 6–7 December, when the Sun was below the horizon from France.

---

* Here, and throughout this book, we are using the New Style calendar. Thus 6 December new style is equivalent to 27 November old style.

Kepler had predicted no more transits of Venus before 1761; he had missed the fact that they occur in pairs, separated by eight years. And this is where Jeremiah Horrocks comes into the story.

Horrocks was born at Toxteth, near Liverpool, in 1619. Not much is known about his early years, but clearly he showed great ability, and by 1632 he had entered Emmanuel College, Cambridge. It was here that he became seriously interested in astronomy, and he made some notable friends, either by personal contact or by correspondence; among them were William Crabtree, of whom more anon, and William Gascoigne, inventor of the micrometer, who was later to "lose his life during the Battle of Marston Moor between Charles I's Cavaliers and Oliver Cromwell's Roundheads". Horrocks was emerging as a mathematician of the first rank, and late in October 1639 he made a careful examination of Kepler's calculations for transits of Venus. He found something very interesting: there might be a transit in December of that year. He was by no means certain – there was a good chance that Venus might pass clear of the Sun – but it was worth following up. Time was short, and all Horrocks could do was to send messages to his brother Jonas, in Toxteth, and to Crabtree, who was living in Salford. Postal services in those days were even slower than they are now, and it is unlikely that anyone else had the slightest inkling that a transit might occur.

Figure 5.1. Stained glass memorial to Horrocks in Hode Church.

Details of Horrocks' career around this time are sparse. It is usually said that he had entered the Church, and had become curate at a church in Hoole, a village fifteen miles from Liverpool, so that in most books he is referred to as "the Rev". In fact, according to Allan Chapman, Britain's leading astronomical historian, there is no evidence that he was ever ordained, and a good deal of indirect evidence that he was not.[*] In 1639 he was only 20 years old, too young to have been a deacon and most certainly far too young to be a full priest. He had the benefit of a Cambridge education, and may have supported himself by teaching. At any rate, at the time of the transit he was certainly in Hoole, and associated in some way with the local church, but by the summer of 1640 he was back home in Toxteth with his family, who had no clerical connections.

Horrocks made what preparations he could, and there is no doubt that he observed the transit from an upstairs window in Carr House, in Hoole, close to the church.[**] His telescope was modest by modern standards; apparently it was a Galilean refractor, but its precise size is not known, and unfortunately Horrocks does not tell us. Sensibly, he used the telescope to project the Sun's image on to a graduated screen fixed behind the eyepiece, as Gassendi had done when following the transit of Mercury eight years earlier; the image on Horrocks' screen was 6 inches across. He did claim that in this way he was able to see "the smallest

---

[*] *Quarterly Journal of the Royal Astronomical Society,* Vol. 31, 133.

[**] In June 1969, at the 200th anniversary of captain Cook's observations of the 1769 transit, I went to Carr House to present a BBC television *Sky at Night* programme. We went to Hoole Church, where there is a stained-glass window in honour of Horrocks (so that he must have had strong connections with the church itself) and then into Carr House, which has been turned into a Barry Elder Doll Museum. *Sky at Night* programmes always open with the same piece of music, "*At the Castle Gate*", from Sibelius' suite *Pelleas and Melisande*. Half an hour before we were due on the air (live) the producer realised that the record had been forgotten. Panic! We looked round for a piano; we did not find one, but there was an ancient harpsichord which worked. I had never played a harpsichord, and I had never played "*At the Castle Gate*". Somehow or other I managed to rehearse, and play us in and out. I wonder what Sibelius would have thought? – P.M.

spots on the Sun", and during the transit he noted that Venus was much blacker than any sunspot could be.

It was natural for him to be cautious, and he began observing well ahead of time. He watched the Sun for much of the day before the predicted transit, and saw nothing apart from a few spots. Luckily the next day was clear, and Horrocks was able to take full advantage of it. It happened to be a Sunday, and the fact that he could spend hours at the telescope is in itself enough to prove that he could not have been a full-time curate.

His observations were described in the only book he was to write, *Venus in Sole Visa*. It was translated into English by the Rev. A.B. Whatton in 1859, and the following quotes are from this translation:

> The chance of a clouded atmosphere caused me much anxiety; for Jupiter and Mercury were in

Figure 5.2. Horrocks observing, recorded in another stained glass window of Hoole Church.

conjunction with the Sun almost at the same time as Venus. This remarkable assemblage of the planets (as if they were desirous of beholding, in common with ourselves, the wonders of the heavens, and of adding to the splendour of the scene), seemed to forebode great severity of weather. Mercury, whose conjunction with the Sun is invariably attended with storm and tempest, was especially to be feared. In this apprehension I coincide with the opinion of the astrologers, because it is confirmed by experience; but in other respects I cannot help despising their more puerile vanities...

Having attentively examined Venus with my instrument, I described on a sheet of paper a circle whose diameter was nearly equal to six inches, the narrowness of the apartment not permitting me conveniently to use a larger size. This however admitted of a sufficiently accurate division; nor could the arc of a quadrant be apportioned more exactly, even with a radius of fifty feet, which is as great an one as any astronomer has divided; and it is in my opinion far more convenient than a larger, for although it represents the Sun's image less, yet it depicts if more clearly and steadily.

...When the time of the observation approached, I retired to my apartment, and having closed the windows against the light, I directed my telescope, previously adjusted to a focus, through the aperture towards the Sun and received his rays at right angles upon the paper already mentioned. The Sun's image exactly filled the circle, and I watched carefully and unceasingly for any dark body that might enter upon the disk of light.

Although the corrected computation of Venus' motions which I had before prepared, and on the accuracy of which I implicitly relied, forbade me to expect anything before three o'clock in the afternoon of the 24th; yet since, according to the calculations of most astronomers, the conjunction should take place sooner, by some even on the 23rd, I was unwilling to depend entirely on my own opinion which was not sufficiently confirmed, lest by too much self-confidence I might endanger the observation. Anxiously intent, therefore, on the undertaking through the greater part of the 23rd, and the whole of the 24th, I watched carefully on the 24th from sunrise to nine o'clock, and from a little before ten until noon, and at one in the afternoon, being called away in the intervals by business of the highest import-ance which, for these ornamental pursuits, I could not with propriety neglect. But during all this time I was nothing in the Sun except a small and common spot, consisting as it were of three points at a distance from the centre towards the left, which I noticed on the preceding and follow-ing days. This evidently had nothing to do with Venus. About fifteen minutes past three in the afternoon, when I was again at liberty to continue my labours, the clouds, as if by divine interposi-tion, were entirely dispersed, and I was once more invited to the grateful task of repeating my observations. I then beheld a most agreeable spectacle, the object of my sanguine wishes, a spot of unusual magnitude and of a perfectly cir-cular shape, which had already fully entered upon the Sun's disk on the left, so that the limbs of the Sun and Venus were precisely coincided, forming an angle of contact. Not doubting that this was really the shadow of the planet, I imme-diately applied myself sedulously to observe it.

In the first place, with respect to the inclina-tion, the line of the diameter of the circle being perpendicular to the horizon, although its place was somewhat inclined on account of the Sun's altitude, I found that the shadow of Venus at the aforesaid hour, namely fifteen minutes past three, had entered the Sun's disk about 62° 30′, certainly between 60° and 65°, from the top towards the right. This was the appearance in the dark apartment; therefore out of doors beneath the open sky, according to the laws of optics, the contrary would be the case, and Venus would be below the centre of the Sun, distant 62° 30′ from the lower limb, or the nadir, as the Arabians term it. The inclination remained to all appearance the same until sunset, when the observation was concluded.

In the second place, the distance between the centres of Venus and the Sun I found, by three observations, to be as follows:

At 3.15 by the clock .... 14′24″

At 3.35 by the clock ……… 13′30″
At 3.45 by the clock ……… 13′0″
At 3.50 the apparent sunset.

The true setting being 3.45, and the apparent about five minutes later, the difference being caused by refraction. The clock therefore was sufficiently correct.

In the third place, I found after careful and repeated observation, that the diameter of Venus, as her shadow was depicted on the paper, was larger indeed than the thirtieth part of the solar diameter, though not more so than the sixth, or at the utmost the fifth, of such a part. Therefore, let the diameter of the Sun be to the diameter of Venus as 30′ to 1′12″. Certainly her diameter never equalled 1′30″, scarcely perhaps 1′20″, and this was evident as well when the planet was near the Sun's limb, as when far distant from it.

This observation was made in an obscure village where I have long been in the habit of observing, about fifteen miles to the north of Liverpool, the latitude of which I believe to be 53°20′; although by the common maps it is stated at 54°12′, therefore the latitude of the village will be 53°35′, and the longitude of both 22°30′ from the Fortunate Islands, now called the Canaries. This is 14°15′ to the west of Uraniburg in Denmark, the longitude of which is stated by Brahe, a native of the place, to be 36°45′ from these Islands.

This is all I could observe respecting this celebrated conjunction, during the short time the Sun remained in the horizon; for although Venus continued on the disk for several hours, she was not visible to me longer than half-an-hour, on account of his so quickly setting. Nevertheless, all the observations which could possibly be made in so short a time, I was enabled, by Divine Providence, to complete so effectually that I could scarcely have wished for a more extended period. The inclination was the only point upon which I failed to attain the utmost precision; for, owing to the rapid motion of the Sun, it was difficult to observe with certainty to a single degree, and I frankly confess, that I neither did not nor could ascertain it. But all the rest is sufficiently accurate, and as exact as I could desire.

So far as we know, Jonas Horrocks – Jeremiah's brother – either missed the transit or else left no record of it. Crabtree had mixed fortunes. For most of the day the sky was completely overcast, and he had almost given up all hope when, just before sunset, there was a break in the clouds. At once he went to his telescope, and projected the Sun's image on to the screen – and there was Venus. According to Horrocks, he was so captivated that he spent some time in simply staring at it, and before he could make any proper measurement the clouds had returned. However, he did make one sketch, and from this it seems that his estimate of the planet's apparent diameter agreed quite well with that of Horrocks; 1′3″ as against 1′12″.

In Manchester City Hall there is a painting of Crabtree observing the transit. Alas, it does not seem to be authentic – but at least it is a superb picture!

The prediction and observation of the 1639 transit was by no means Horrocks' only achievement. He seemed destined for a brilliant career, but it was not to be. Soon after the transit he sat down to write *Venus in Sole Visa,* and when looking round for a publisher he communicated with Crabtree, suggesting that the two should meet to discuss it. Crabtree was only too eager, and they planned to meet on 2 January 1641. They never did; on the previous day Horrocks died suddenly. It was a tragic end to a brief but amazingly productive life. Crabtree survived him by only three years; he died in late July 1644. Yet the work of these two young men will never be forgotten, and they alone were the observers of the first transit of Venus to be both predicted and confirmed.

Figure 5.3. Crabtree observing transit of Venus, Manchester City Hall

Chapter 6

The Transit of 1761

There can be little doubt that the moving spirit behind the transit observations of 1761 and 1769 was Edmond Halley, even though he had been dead for almost twenty years before the first transit of the pair took place. He originally planned to use transits of Mercury to help in measuring the length of the astronomical unit, but Venus would be a much more satisfactory target, partly because it is closer and also because it is larger. As long ago as 1716 Halley had delivered a paper to the Royal Society, from which I quote:

> We therefore recommend again and again, to the curious investigators of the stars to whom, when our lives are over, these observations are entrusted, that they, mindful of our advice, apply themselves to the undertaking of these observations vigorously. And for them we desire and pray for all good luck, especially that they may not be deprived of this coveted spectacle by the unfortunate obscuration of cloudy heavens, and that the immensities of the celestial spheres, compelled to more precise boundaries, may at last yield to their glory and eternal fame.

In fact the transit was observed from over a hundred and twenty sites, and by almost two hundred astronomers. To list them all would be out of place here, but it seems helpful to tabulate some of the more important locations, arranged here in order of longitude from west to east (see Table 6.1). The telescopes used are

Table 6.1. Observations from the transit of 1761

Site	Observer	Nation	Focal length	Ingress	Egress
Peking	Dollier	France	14	2	1,2
Tobolsk	Chappe	France	19	2	1,2
Isle de Rodrigue	Pingré	France	18	1	–
Cajaneborg	Planman	Sweden	21	2	1,2
Cape of Good Hope	Mason, Dixon	Britain	2*	1,2	–
Stockholm	Wargentin	Sweden	19	2	1,2
Uppsala	Bergmann	Sweden	21	2	1,2
Calmar	Wickström	Sweden	21	2	1,2
Vienna	Hell	Germany	4.5*	–	2
Vienna	Cassini	France	9	–	2
Copenhagen	Horrebow	Denmark	22	–	1,2
Trondheim	Bugge, Hascow	Denmark	8	–	1,2
Rome	Audifredi	Italy	?	–	1,2
Bologna	Zanotti	Italy	2.5	–	1,2
Florence	Ximenes	Italy	4.3*	–	1,2
Göttingen	T. Mayer	Germany	6	–	1,2
Leiden	Luloss	Holland	7*	–	1
Paris	Maraldi	Italy	15	–	1.2
Lalande	Luxemburg	France	18	–	1,2
Passy	Ferrner	Sweden	28*	–	1,2
Rouen	Bouin	France	16	–	2
Greenwich	Bliss	Britain	15	–	1.2
London	Short	Britain	2*	–	1,2
Oxford	Hornsby	Britain	12	–	1
Madrid	Rieger	Germany	8	–	1.2
Porto	De Almeida	Portugal	2*	–	1,2
Constantinople	Porter	Britain	1.5*	–	1,2
Winthrop	St. John's, Newfoundland	Britain		–	1,2

classified according to their focal lengths in feet – in some cases the apertures are not known. In the Ingress and Egress columns, 1 and 2 indicate the inner and outer contact respectively. Telescopes are refractors, except those marked *, which are reflectors.

Not listed here is the important observation made by Mikhail Lomonosov, the first of Russia's great astronomers. Observing from St. Petersburg, he found that the outline of Venus was blurred rather than distinct, and he put this down to the presence of an atmosphere. Indeed, he wrote that Venus "is surrounded by a considerable atmosphere, equal to, if not greater than, that which envelops our earthly sphere". Rather surprisingly, this attracted little attention at the time, and the famous French astronomer Lalande still maintained that Venus' atmosphere was much thinner than that of

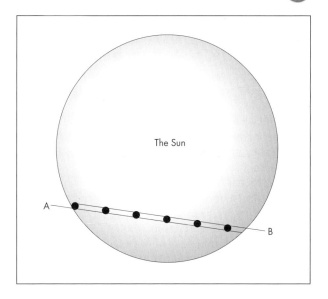

Figure 6.1. The conditions for two observers in 1761. A: Chappe was in Tobolsk. B: Pingré was on the Isle de Rodrigue.

the Earth. We now know, of course, that Lomonosov was quite right.

Remember that Halley's method of measuring the astronomical unit depended upon observations at both ingress and egress – and the 1761 transit was not suited to this; that of 1769 was expected to be much better. Still, at least 1761 provided a useful rehearsal, and some very valuable observations were made quite apart from Lomonosov's.

Yet the final results were much less precise than Halley had hoped. The problem arose from a phenomenon which Halley did not foresee, and could not possibly be expected to have foreseen. As Venus moves on to the Sun it seems to draw a strip of blackness after it; the entire shape is distorted. When the strip disappears the transit has already begun, so that giving an exact timing is well nigh impossible. This "Black Drop" is due to the atmosphere of Venus, and there is no way of eliminating it; it was equally obtrusive at the transits of the 19th century – and will no doubt be again in 2004 and 2012.

There was another problem, too, which affected European observers. Travel to far climes was essential – conditions in the West were unsuitable – and England and France were busy at one of their interminable wars. Several expeditions were badly hampered, and one at least was totally ruined.

The First French Expedition

Let us begin with France, where the main instigator was Joseph-Nicolas Delisle. Delisle was born in 1688, and for a while became an observer in Luxembourg; in 1724 he visited England and actually had discussions with Halley about the idea of using transits to measure the solar parallax. He was then invited to Russia, and between 1725 and 1747 acted as Director of the St. Petersburg observatory; subsequently he returned to France. As we have noted, his method of using the transits was different from Halley's, and involved nothing more than timing of ingress.

Initially Delisle hoped that transits of Mercury could be used, but the results of the 1753 transit were disappointing, and he had to accept that Venus provided much more opportunity. He began to organise expeditions for 1761 and 1769 (though in fact he died a year before the second transit of the pair took place). The three most interesting of the French expeditions were those of Chappe, Pingré and Legentil, scheduled to go to Siberia, Rodrigues in the Indian Ocean, and Pondicherry in India respectively. Both the latter sites were then under French control.

The Abbé Jean Chappe d'Auteroche was an ideal choice as leader of one major expedition. (He is always referred to as "Abbé", though whether or not he was actually ordained seems rather uncertain.) He was born in 1727, and joined the staff of the Paris Observatory, initially as a draftsman; his ability soon became evident, and he was also very popular personally. He observed the 1753 transit of Mercury, together with the Director of the Paris Observatory (Cassini de Thury) and two other eminent astronomers, Maraldi and Legentil; in 1760 he became an official astronomer at Paris, and Cassini had no hesitation in recommending him to play a leading rôle in the 1761 transit of Venus.

The site selected for Chappe was Tobolsk, in Siberia. It is not a particularly accessible place even now, and was even less so in the 18th century, so that Chappe and his team had to be ready for a long and difficult journey. So it proved. He set out in November 1760, and went first to Holland to go on board a ship for St. Petersburg. When he arrived, the ship had already departed – which in the event proved to be just as well, because it went aground off the Swedish coast. Chappe decided to go by land, but the roads were poor, and once a carriage carrying much of the equipment fell into a ditch, though fortunately the main instruments were not damaged. They reached Vienna, and there met the famous astronomer Maximilian Hell, of whom more anon. Warsaw was the next stop, and then St. Petersburg; he finally reached Tobolsk on 10 April – not a moment too soon, because in another week the summer thaw would have made travel impossible; river ice served as sturdy bridges for the sledges.

By the end of the second week in May Chappe had erected a wooden observatory, and equipment was unpacked and tested; a partial solar eclipse on 3 June showed that everything was in order. There was, however, one factor which Chappe had not taken into account. The local inhabitants became convinced that he was a magician, intent on destroying them, and by bad luck the main river in the area had flooded the countryside. To ignore such a threat would have been unwise in the extreme. Chappe recorded that "some of the Russians advised me not to go alone to the observatory, and to take some precautions against the fury of the mob, which might lead them to any lengths... From that time I determined to pass most nights in the observatory, lest they should attempt to pull it down. On the fourth instant, the wind was so high and so continual, that I was in the gravest danger of having it blown down, and was not freed from my apprehensions until the next day at noon, when the wind abated."

On the night before the transit the sky became totally overcast. Chappe's diary describes his feelings: "The idea of returning to France, after a fruitless voyage; of having exposed myself in vain to a variety of danger, and to fatigues, under which I was supported only by earnestness and expectation of success, which I was now deprived of by a cloud, at a time when I had the greatest reason to be assured of it, threw me into such a situation as can only be felt... In these dreadful agitations I passed the whole night; I went out and came in again every instant, and could not continue a moment in the same position."

Fate was kind. Chappe again: "My watchmaker's business was to write, and keep eye on the clock, while my interpreter was employed in counting the time; the calmness and serenity of the air had made me resolve to bring the instruments out of the observatory, that might move them more readily. I soon perceived one of the borders of the Sun, at the time when Venus was to enter upon his disk; but on the opposite border, which was still concealed by clouds. I stood fixed with my eye to the telescope, wandering over the immense space between us and the Sun a thousand times a minute. I was troubled by the continuance of the cloud, which at length disappeared, and perceiving that the planet was already immersed, I prepared to observe the most material appearance, the total entry. Although the sky was perfectly serene, yet my apprehensions were not yet at an end. The moment of observation was now at hand; I was seized with a universal shivering, and was obliged to collect all my thoughts, in order not to miss it. At length I observed this phasis, and felt an inward persuasion of the accuracy of my process. Pleasure of the like nature may sometimes be experienced, but at this instant, I truly enjoyed that of my observation, and was delighted with the hopes of its being still useful to posterity, when I had quitted this life."

He left Tobolsk on 28 August, and after a pleasant stay in St. Petersburg sailed for France, arriving back after an absence of two years. His report was concise and accurate; he published three large volumes entitled *Voyage en Sibérie*, giving a detailed account of the whole expedition.

The Second French Expedition

The second French expedition was led by Alexandre-Gui Pingré, who began his career as a professor of theology and then became an astronomer at the Academy of Sciences in Rouen. His assistant during the transit observations was Denis Thuillier, a competent astronomer though apparently he and Pingré were by no means close friends. They sailed on the good ship Comte d'Argenson in January 1761, and finally reached their destination. Rodrigue island in the Indian Ocean, on 28 May – not without incident; the Anglo-French war was in full swing, and at one time it was even proposed to divert the Comte d'Argenson in order to go to the help of a French ship which had been damaged in battle. They arrived with little time to spare, and Rodrigue, an island of basaltic rock rising steeply from the sea, was an awkward site; somehow they managed, and set up their equipment. The ingress was observed, but then the clouds came up and nothing more could be done at least from a purely astronomical point of view.

However, they did not leave immediately; they made a series of valuable latitude and longitude determinations, and also made biological studies. Unfortunately the island peace was soon to be shattered.

In late June the French corvette l'Oiseau arrived, followed three days later by the British warship Plassey, which proceeded to attack. The island was captured, and the l'Oiseau burned. The French officers and seamen were left on the island together with Pingré and Thuillier – and there they stayed until September, when a passing French ship happened to find them and take them to the Isle de France (Mauritius), which was very much less isolated.

From there they went on board the frigate Boutin, and made their way to another island, the Isle de Bourbon, where they spent eight weeks exploring the area and making biological observations. Finally they departed for home. It was too much to hope that they would be left in peace – and, sure enough, they encountered another British warship. After a brief exchange of shots the French captain prudently surrendered, and the Boutin was boarded. Some of the crew were taken prisoner, and the passengers' belongings raided, but this time the astronomers escaped lightly, and were allowed to stay on the Boutin. Eventually they reached Lisbon, and proceeded to overland to Paris, reaching home at the end of May.

All in all, it may be said that Pingré's expedition was a qualified success. True, the observations were curtailed by cloud, but some valuable results were obtained, and the two astronomers did their best under very difficult circumstances.

The Third French Expedition

Next we come to Legentil – or, to give him his full name, Guillaume Joseph Hyacinthe Jean Baptiste le Gentil de la Galaisière, whose expedition must surely have been the unluckiest in the whole of astronomical history.

Legentil was a thoroughly competent astronomer, who worked at the Paris Observatory with Cassini de Thury; he was, incidentally, one of the first to realise that transits of Mercury could never be observed with sufficient accuracy to make them really valuable in measuring the length of the astronomical unit. In 1760 he announced that he proposed to go to Pondicherry in India and observe the transit from there, which was sensible enough.

Legentil sailed from Brest on 16 March 1760, taking full equipment with him. The first leg of his journey, to the Isle de France, was uneventful apart from a near-encounter with British ships of the Cape of Good Hope, but then came bad news. Pondicherry was under siege, and there had also been a disastrous hurricane. Legentil changed his plans. It seems that for a while he considered going to Batavia, but then came orders from Paris; French soldiers were to be sent to Pondicherry. Legentil took the opportunity to join them, and went on board the La Sylphide. Before they could arrive, Pondicherry had fallen. There was no choice but to turn round and head back to the Isle de France. Legentil had a perfect view of the transit – but from the deck of the ship, and he could make no scientific observations whatsoever.

What next? Rather than return home empty-handed, he elected to remain in the East until 1769, and observe the second transit instead. He was not idle, and in fact he made a great many varied and valuable scientific observations, but the transit was always uppermost in his mind, and he gave careful consideration to the choice of site. He finally decided upon Manila, in the Philippines, and arrived there in August 1766, which certainly gave him plenty of time to prepare.

Then, in July 1767, Legentil received a letter from France. The French authorities were very anxious for him to go back to Pondicherry rather than stay in the Philippines. Finally Legentil agreed, possibly because he had run foul of the corrupt Governor of Manila, and by March 1768 he was back in Pondicherry. This time he had full co-operation from the British, who even loaned him an excellent 3-foot telescope, and once again he set up his equipment.

On 2 June 1769, the day before the transit, the weather was perfect. Then the clouds rolled in; Legentil had to wait under an overcast sky, praying for a miracle which did not happen. Half an hour after the end of the transit, the sky was crystal-clear again... And it then transpired that at Manila there had been no cloud cover at all.

One can well imagine Legentil's feelings. It was rather too long to wait for the next transit (that of 1874), so at last he decided to go home. After two minor shipwrecks he reached Paris in October 1771, to find that he had been presumed dead and that his heirs were preparing to distribute his property; he took legal action to reclaim it at something later than the eleventh hour. At least it is pleasant to record that his later life was happy; he married, and his wife and daughter "made him forget the misfortunes of the past", according to Cassini. He lived on until 1792.

The British Expeditions

British preparations started rather later than the French, but – perhaps prompted by news of Delisle's work! – the Royal Society decided to dispatch as many expeditions as possible. Nevil Maskelyne, later to become Astronomer Royal, went to the island of St. Helena, accompanied by his assistant Robert Waddington; alas, they were clouded out. Even though they had taken the best available equipment, and had obtained a generous grant of £800 (!) for the expedition, they were unable to do any astronomical work at all. Occasionally they were able to glimpse Venus against the solar disk, but conditions were too bad for them to make any useful measurements, and – through no fault of their own – the St. Helena foray was a complete failure.

Professor John Winthrop, who took a party as far afield as St. John's in Newfoundland, had much better luck. They arrived in good time, set up their equipment and made their observations under excellent conditions. Indeed, theirs were some of the best results obtained in 1761.

Another expedition, organised by the Royal Society, was to be sent to Bencoola in the East Indies under Charles Mason and Jeremiah Dixon. Because of the war situation, they set out in a naval ship, HMS Sea Horse, which was promptly attacked by the French frigate Le Grand and had to turn back to Plymouth – not without suffering casualties (at least eleven sailors were killed during the battle). Mason and Dixon were unhurt, but they were thoroughly unnerved, and told the Royal Society that "we will not proceed thither, let the consequences be what it will". The Royal Society sent a stinging reply; their refusal would be "a reproach to the

Nation in general, to the Royal Society in particular, and fatally to themselves..." Mason and Dixon capitulated, and set off again, but clearly they could not hope to get to Bencoola in time, so they stopped off at the Cape of Good Hope and observed from there – successfully, it must be added. It was just as well that they did not go on to Bencoola, because by that time it had fallen into French hands.

Other Observations

Good observations were made by Per Wargentin, from Stockholm, who was probably the first to define the problems caused by the "Black Drop"; Cassini de Thury went to Vienna and made some measurements of the egress; even at Greenwich Bliss, briefly Astronomer Royal, was able to see both contacts of the egress. All things considered, the 1761 transit was certainly the most elaborate international programme in astronomy ever mounted up to that time. And yet, in the end, it has to be admitted that the final results were disappointing. The observations showed a great deal of scatter, and the solar parallax determinations ranged between 8.5 seconds of arc, and 10.5 seconds of arc, corresponding to distances of 96 162 840 miles and only 77 846 110 miles. In 1822 Johann Encke, a very famous German astronomer, analysed all the observations and gave a parallax from 8.429813 seconds to 8.551237 seconds, corresponding to distances between 97 000 000 miles and 95 600 000 miles.

Where did it go wrong? There were two main reasons. One was that most of the observations depended upon Delisle's method, and the latitudes and longitudes of the sites were not known with sufficient precision; Halley's method involved timing both ingress and egress, and as can be seen from Table 6.1 on page 38 there were not many observations of ingress in 1761. Secondly, of course, there was the dreaded "Black Drop".

Yet despite the uncertainties, it had been a useful exercise. Astronomers waited impatiently for the transit of 1769.

Chapter 7

Captain Cook and the Transit of 1769

Several important lessons had been learned from the transit of 1761. There was nothing to be done about the "Black Drop", but at least it had become clear that if a really good value for the astronomical unit were to be obtained it would be necessary to send expeditions all over the globe – as far north and as far south as possible. In the end over 150 sets of measurements were made from over 80 stations, and the overall programme as truly international, as can be seen from Table 7.1.

This time Britain did not intend to let the opportunity slip, and the initiative was taken by Thomas

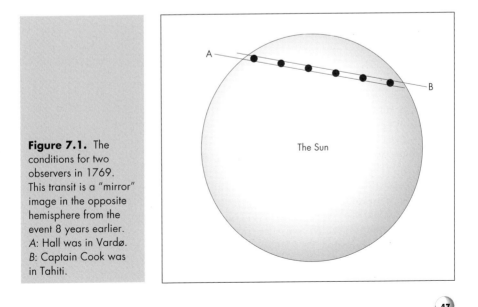

Figure 7.1. The conditions for two observers in 1769. This transit is a "mirror" image in the opposite hemisphere from the event 8 years earlier. *A*: Hall was in Vardø. *B*: Captain Cook was in Tahiti.

The Sun

Table 7.1. Observations from the transit of 1769

Site	Observer	Nation	Focal length	Ingress	Egress
Yakutsk	Islenieff	Russia	10	2	1.2
Manila	De Ronas	Spain	15	1,2	1,2
Peking	Dollières	France	18	–	1,2
Orenburg	Krafft	Russia	12	–	1,2
Vardø	Hell	Denmark	10*	1,2	1,2
St Petersburg	C Mayer	Russia	18	–	2
Cajaneborg	Planmann	Sweden	12	2	2
North Cape, Norway	Bayley	Britain	22*	2	–
Hammerfest	Dixon	Britain	2*	2	–
Stockholm	Wargentin	Sweden	21	1,2	–
Stockholm	Ferrner	Sweden	10	1,2	–
Uppsala	Prosperin	Sweden	16	1,2	–
Griefswald	A. Mayer	Germany	7	1.2	–
Paris	Messier	France	12*	2	–
Paris	Cassini de Thury	France	3.5*	2	–
St Hubert	Le Monnier	France	10*	2	–
Greenwich	Maskelyne	Britain	2*	1,2	–
Greenwich	Horsley	Britain	10*	1,2	–
Oxford	Hornby	Britain	7.5*	1,2	–
Kew	Bevis	Britain	3.5*	1,2	–
Lizard Point	Bradley	Britain	2*	1,2	–
Cavan, Ireland	Mason	Britain	2*	1,2	–
Gibraltar	Jardine	Britain	2*	1,2	–
Cambridge, Mass., USA	Winthrop	Britain	2*	1,2	–
Philadelphia	Shippen	Britain	2*	1,2	–
Norristown	Lukens	Britain	41*	1,2	–
Hudson's Bay	Dymond, Wales	Britain	2*	2	1,2
San José, Baja California	Chappe	France	3*	2	1,2
St Anne, California	Velázquez	Spain	?	2	1,2
Green	Tahiti	Britain	2*	1,2	1,2
Cook	Tahiti	Britain	2*	1,2	1,2
Solander	Tahiti	Britain	3*	1,2	1,2

Hornsby, Savilian Professor of Astronomy at Oxford. In 1767, he wrote to the Royal Society: "Astronomers of the present age are peculiarly fortunate in being able so soon to have recourse to another transit of Venus in 1769, when, on account of the planet's north latitude, a difference in the total duration may conveniently be observed, greater than could possibly be obtained, or even expected, by Dr Halley, from the last transit." Calculations (made chiefly by Pingré, in France) showed that from north Norway the transit would last for 5 hours 55 minutes 10 seconds, while form the South Pacific the duration would be 5 hours 26 minutes 36 seconds – a difference of over 28.5 minutes.

In June 1766 the Council of the Royal Society met to make preliminary plans. Hornsby had recommended that an observing site should be selected in the South Seas, or in Mexico or Baja California, which were then under Spanish control. The Royal Society Council inclined to Baja California, and promptly invited a distinguished foreign astronomer to lead the party – Tudher Josip Boscovic, better known as Father Roger Boscovich. The choice was a good one. Boscovic had a fine reputation; he was Professor of Mathematics at Pavia University, and was a foreign Fellow of the Royal Society. He accepted at once, but there were immediate problems. Boscovich was a Jesuit, and around this time the Jesuit Order was suppressed, so that under no conditions would Boscovich be allowed to go to Spanish America. The Council then asked whether an expedition would be allowed to go to California provided that Boscovich did not accompany it. Again the answer was "no", because at that time the British and Spanish governments were on the very worst of terms. At this juncture the Royal Society abandoned all thought of Baja California, and turned its attention elsewhere.

Figure 7.2. The atmosphere of Venus seen during a transit.

The Spanish authorities were much more friendly toward France, and the French decided to send their own expedition to Baja California. Who better to lead it than Chappe, who had been so efficient in 1761?

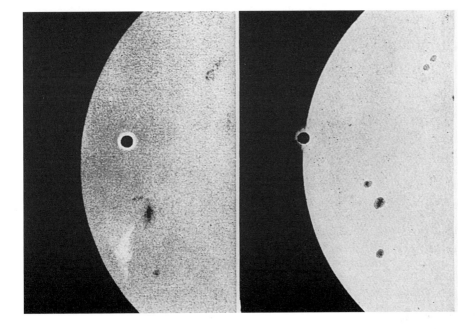

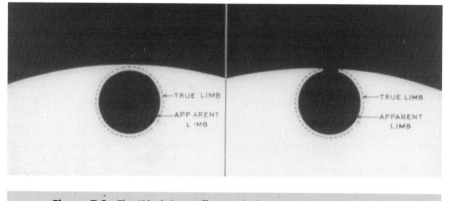

Figure 7.3. The "Black Drop"effect partly due to the planet's atmosphere.

Chappe was only too glad to go. The only restrictions placed upon him were that he must travel in company with a Spanish team, and that he would do no more than carry out a purely scientific programme (which was all he wanted to do in any case).

To go with him, Chappe chose "Mr Pauly, the King's Engineer and Geographer, from whose talents I expect great assistance, who was to second me in my astronomical and geographical observations; Mr Noël, a pupil of the Academy of Painting, who was intended for our draftsman, to take drafts of sea coasts, plants, animals, and whatever we might meet with that was curious; lastly Mr Dubois, a watchmaker, who was instructed with the care of preserving my instruments, and repairing little mischiefs they too often sustain during long voyages." The equipment included a 3-foot quadrant, and 18-inch English quadrant, two telescopes of 10- and 3-feet focal length respectively, a transit instrument, and a pendulum clock.

Chappe left Paris on 18 September 1768, sailed from Le Havre three days later, and reached Cadiz in mid-October. From there the party sailed to Vera Cruz, on the Gulf of Mexico, where they were joined by two Spaniards, Vicente de Doz – a competent astronomer – and Salvador de Medina, about whom not much is known but who also was presumably well-informed. They made their way overland to Mexico City (the astronomers carried in litters, the other members of the party on muleback) and then made ready to cross to San Blas, on the Mexican coast. On arrival there they were well received. Chappe wrote: "I am at a loss for words to express the friendship and politeness shown to us by the Marquès de Croix, Viceroy of Mexico, and

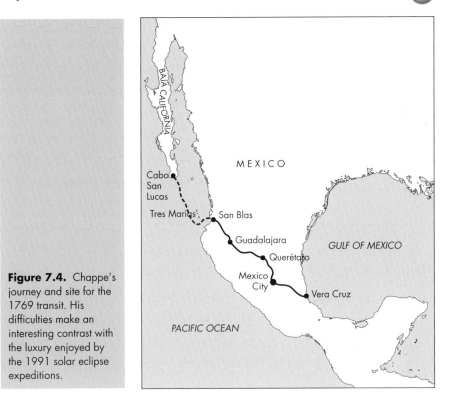

Figure 7.4. Chappe's journey and site for the 1769 transit. His difficulties make an interesting contrast with the luxury enjoyed by the 1991 solar eclipse expeditions.

by his whole court... We had no table but his own for the four days we continued in the town, and he was so obliging as to send us a cook to dress victuals for our attendants after the French fashion."

They had to endure a sea crossing to the actual observing site, San José del Cabo in Baja California, and it was a rough passage; for a while it was doubtful whether a landing would be possible at all. Both Doz and Medina were apprehensive, but Chappe persisted, and eventually they managed to transport themselves and their baggage to dry land. Fortunately the Spanish "visitor general" to the area, Jose de Galvaz, proved to be as co-operative as the Marquès had been, though he had to leave for home before the day of the transit and never actually met Chappe's party.

Chappe set up a makeshift observatory, while Doz and Medina erected another observatory close by. The weather was set fair, but there was one terrible threat: disease. What was called "epidemical distemper" was raging in the area, and had already killed at least one-third of the population. It seems to have been a form of typhus, and of course there were no preventive measures available to combat it.

The prudent course would have been to abandon the expedition and leave, but – rather against the advice of his colleagues – Chappe elected to stay. Another observer was also ready: Don Joaquin Velázques Cardenas de Leon, administrator of the local mining district, who was a keen and skilful astronomer. His chosen site was just over fifty miles from Chappe's, and he too was favoured with good weather on June the third.

All the observations – those by Chappe, Doz, Medina and Velázquez – were successful, and were possibly the best made anywhere during the 1769 transit. Chappe took the actual observations, while a servant counted out the seconds, Pauly recorded the minutes and took notes, and Dubois made sure that the instruments were working perfectly. Doz and Medina worked together, and their results closely matched those of Chappe; so too did those of Velázques.

But disaster lay ahead. Three days after the transit Doz and Medina were taken ill, and so were many of the Spanish helpers. Chappe did his best to help, but then he too was stricken. He fought the disease as long as he could, and even managed to observe an eclipse of the Moon on 18 June, but his condition became steadily worse, and on 1 August he died. Dubois survived him by only a few days, and Medina died on the return journey. Only Pauly and Doz lived to return home (Doz went on to a distinguished naval career, rising to the rank of rear-admiral).

It is a tragic story, unique in astronomical history. Chappe could have saved himself; but for his devotion to duty he could have left the area in time to avoid infection. At least he had the satisfaction of knowing that his work had been well done, and that the results would be of immense value to his fellow scientists.

One expedition which caused a good deal of controversy was that of Father Maximilian Hell, who had been born in Hungary in 1720 and had made a reputation as a skilful astronomer. In 1755 the Empress of the Austro-Hungarian Empire made him court Astronomer, and he became director of the Vienna Observatory, where he carried out much excellent work.[*]

In 1767 King Christian VII of Denmark and Norway invited Hell to observe the 1769 transit from Vardø, an

---

[*] A lunar crater is named Hell in his honour. This does not, however, indicate that the crater is exceptionally deep!

Arctic island of the coast of Lapland. Hell accepted, and observed the transit successfully, but he did not return home at once, he stayed on Vardø for the best part of a year, carrying out various scientific observations in fields ranging from biology to natural history, meteorology and oceanography. This meant that he delayed in publishing his results, and the French began to wonder if he meant to hold them up until he could be sure that they fitted in with measurements made elsewhere. Hell's work finally came out in 1772. Much later, in 1836, Karl von Littrow – who became director of the Vienna Observatory – examined Hell's diaries and came to the startling conclusion that they had been faked; some entries had been crossed out and rewritten in ink of a different colour. Von Littrow published his results, and Hell's reputation was ruined.

Then, in 1883, the diaries were re-examined by Simon Newcomb, one of America's most eminent astronomers. Newcomb proved that Hell's observations were perfectly genuine; the alterations had been made at the time, and different ink had had to be used in the Arctic chill. Von Littrow had jumped to his conclusions for a curious reason: he was colour-blind, and could not even tell a red star from a white one. Hell's reputation was vindicated. In a letter to Newcomb, John Hagen, of Georgetown Observatory – like Hell, a Jesuit – wrote: "It was fitting that his act of justice should be reserved to an American astronomer who stands aloof from the petty quarrels of the old world."

Despite abandoning California, British interest was as great as ever. There were careful preparations; each expedition was equipped with a quadrant, a clock, two 2-foot reflecting telescopes with micrometer attachments, a barometer, a compass and a thermometer. Eventually William Wales and Joseph Dymond went to Hudson's Bay, William Bayley to North Cape and James Dixon to Hammerfest in north Norway, while Mason, who insisted on a warmer climate, preferred to remain in Ireland, and Winthrop made his observations from Cambridge in Massachusetts. That left the South Seas – and Captain James Cook.

Actually, Cook was not in the original reckoning at all. The chosen observer was Alexander Dalrymple, who was well qualified for the task and was so an excellent navigator and geographer. Funds were provided by the Crown; indeed King George III himself was very interested in astronomy and went so far as to have a small observatory built at Kew so that he could watch the transit personally. (The observatory is still there.) A 370-ton former coal barque was bought and fitted out for the voyage, and Dalrymple assumed that he was to be put in command of it. This was where the first hitch occurred. Dalrymple was quite definite in his views: "I can have no thoughts of undertaking the Voyage as a Passenger going out to make the Observations, or on any other footing, than that of having the management of the Ship intended for the Service." There was, however, an unfortunate precedent. As we have noted, Edmond Halley had been in command of the pink Paramour upon a scientific voyage, and there had been trouble with the officers of the ship's crew – mainly, it must be said, because Halley was a far better navigator than anybody else.

Subsequently the Admiralty had introduced a regulation stating that the commander of a King's ship must also be an officer in the King's Navy. Dalrymple was not. Neither he nor the Admiralty would give way, and Dalrymple withdrew from the whole project in a huff. In retrospect it is rather fortunate that he did, because the Admiralty called in a young, unknown Naval lieutenant to take charge. On 5 May 1768 James Cook was brought before the Royal Society council and appointed captain of the Endeavour.

The senior astronomer was to be Charles Green, who had been an assistant at Greenwich Observatory. Maskelyne knew him and trusted him; the two men had been together on a voyage to Barbados in 1762–63 to carry out the secor test of John Harrison's chronometer, and he was known to be an accurate and conscientious observer. (He was, incidentally, the brother-in-law of William Wales, who went on Cook's second voyage.) Also on board were two scientists: Joseph Banks, who was concerned mainly with natural history, and Daniel Solander, whose interests lay in the same field but who was also a competent astronomical observer. Moreover, Cook himself was quite able to make accurate measurements. Rather as an afterthought, it was decided to continue the voyage after the transit had been observed, since one of George III's greatest ambitions was to play a part in the exploration of unknown parts of the world – even though he had no thought of going on the trip himself. In the event, this proved to be a momentous decision.

The next point to be decided was: where, in the South Seas, should the observers go?

It was obviously essential to fix the precise position, and for once Fate was kind. Shortly before the expedi-

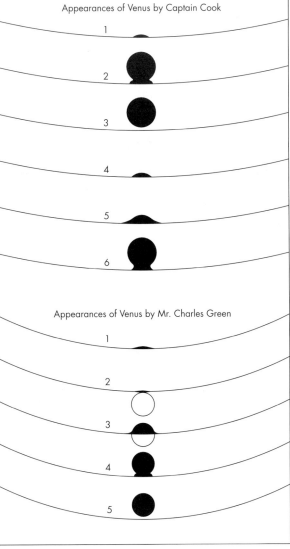

Appearances of Venus by Captain Cook

Appearances of Venus by Mr. Charles Green

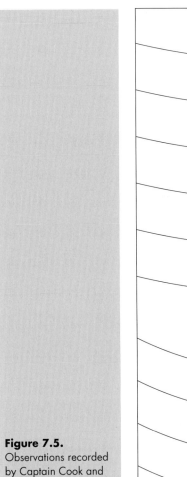

Figure 7.5.
Observations recorded by Captain Cook and Charles Green at the 1769 Venus transit. The "Black Drop" effect seen by the two observers is differently recorded.

tion was due to depart, Captain Samuel Wallis, commander of H.M.S. Dolphin, returned from the Pacific to announce the discovery of the island we know as Tahiti. He reported that the natives seemed friendly – and, more importantly, he had done his best to pinpoint the latitude and longitude. So, on 9 June 1768, the Royal Society notified the Admiralty that Tahiti was to be the chosen site, and the Admiralty did not demur. The die was cast, and on 26 August 1768 the Endeavour sailed from Plymouth, bound for the other side of the world.

The outward voyage was uneventful. Cook and Green made daily observations of position as well as of magnetic variation; it was records of this kind which proved to be so valuable later on when the time came to draw up precise charts. The Endeavour crossed the Line on 25 October, and on 13 November put in at Rio de Janeiro to load up with provisions. (The local envoy was apparently rather wary of Cook and his crew, and recorded that they were on their way south "to observe the transit of Venus, very interesting to the advancement of navigation, of which phenomenon he [the envoy] appeared to be totally ignorant". They then doubled the Horn, "with as much ease as if it had been the North Foreland on the Kentish coast", and by 10 April 1769 they stood off Tahiti. The natives' own name for the island was Otaheite. Europeans had had their first news of it in 1607, when it had been found by Pedro Quitos; when Captain Samuel Wallis rediscovered it in 1707 he named it King George's Island, which is what Cook called it.

The first step was to erect a temporary observatory. As Wallis had reported, the local inhabitants were perfectly friendly, but there was one misadventure which might have proven troublesome. One of the main instruments was a quadrant made by the famous instrument constructor Bird: it had been specially bought, at Cook's request, for the then enormous sum of £13, and when the expedition reached Tahiti the case containing the quadrant was still unopened. It was, naturally, put into Charles Green's tent. The locals, watching eagle-eyed, saw how carefully it was being handled and concluded that it must contain something very precious indeed. When Green went to check the quadrant, he found that both it and the case had disappeared. It would be wrong to suggest that panic ensued, but certainly both Cook and the scientists were alarmed; the quadrant was an essential part of their equipment.

Fortunately it was recovered in good time, but not before it had suffered some rough handling from the inquisitive thief. Much later, after the expedition had returned home, Nevil Maskelyne questioned the accuracy of some of the observations made with it by Charles Green, and Cook complained that the incident should have been remembered .

There have been many cases of scientists who have made long and dangerous journeys in order to observe some phenomenon (such as a total solar eclipse) only to be thwarted by clouds. This is annoying even with an eclipse, but there will always be other chances. With the 1769 transit of Venus this was not so; had the opportunity been missed it would not have recurred until 1874, so that it was literally "now or never". Luckily all went well, and from Tahiti the weather was perfect. Not a single cloud was seen during the entire day, and the atmosphere was calm and steady.

Let Cook tell the story in his own words, as published in his official report contained in the *Philosophical Transactions of the Royal Society,* Vol. 61, page 1379 ff.:

We came to an anchor in Royal Bay in King George's Island.

Fixed upon the North point of the bay, which is the most Northern point of the island, for the place of observation; here we built a small fort, to secure us against the natives, which we called fort Venus; it was not finished and the instruments set up in proper order until the 10th of May, therefore the time for all observations made before this day, was taken by a watch with a second hand, the going of which was ascertained by altitudes of the sun as often as were necessary.

The astronomical clock, made by Shelton and furnished with a gridiron pendulum, was set up in the middle of one end of a large tent, in a frame of wood made for the purpose at Greenwich, fixed firm and as low in the ground as the door of the clock-case would admit, and to prevent its being disturbed by any accident, another framing of wood was made round this, at a distance of one foot from it. The pendulum was adjusted exactly to the same length as it had been at Greenwich. Without the end of the tent facing the clock, and 12 feet from it, stood the observatory, in which were set up the journeyman clock, and astronom- ical quadrant; this last, made by Mr Bird, of one foot radius, stood upon the head of a large cast fixed firm in the ground, and well filled with wet heavy sand. A centinel was placed continually over the tent and the observatory, with order to suffer no one to enter either the one or the other, but those whose business it was. The telescopes made use of on the observations were – Two reflecting ones of two feet focus each, made by the late Mr James Short, one of which was furnished with an object glass micrometer. Thus furnished, the following observations were made.

Cook then gives details of the measurements, including the determination of latitude, which he gave as 17°29′15″ S. and longitude as 149°32′30″ W. The latitude determinations were made between the 2nd and the 28th of June, and depended upon seven sets of measurements. It was these which came in for some criticisms by Maskelyne, who even appended a note in the *Philosophical Transactions*: "It must be confessed, that the results of these observations (most of which were made by Mr Green) differ more from one another than they ought to do, or than those do made by other observers, with quadrants of the same size, and made by the same artist, the cause of which, if not owing to want of care and address by the observer, I don't know how to assign." It is true that the measurements did show some scatter, and Cook was probably justified when he commented that Maskelyne should have taken the thief's treatment of the quadrant into account. However, there is no reason to suppose that the observation of the transit itself were adversely affected.

To give the whole series of transit measurements made by Cook, Green and Solander would be tedious (they are printed in full in the *Philosophical Transactions* report), but again there was some scatter, owing to the awkward behaviour of Venus itself rather than to any instrumental or observational deficiencies. Cook and Green agreed with the timing of the first internal contact of Venus' limb and the Sun, but the Black Drop effect was unpleasantly obtrusive. With the second external contact at the end of the transit, Green's value was 6 seconds later than Cook's and one second later than Solander's. But it is certainly worth quoting Cook's description of the observations, because it does bring out the difficulty of the whole procedure:

> The first appearance of Venus on the Sun, was certainly only the penumbra, and the contact of the limbs did not happen till several seconds after, and then it appeared as in fig. The 4th; this appearance was observed both by Mr Green and me; the time it happened was not noted by either of us; it appeared to be very difficult to judge precisely of the times that the internal contacts of the body of Venus happened, by reason of the darkness of the penumbra at the Sun's limb, it being that nearly, if not quite, as dark as the planet. At this time a faint light, much weaker

than the rest of the penumbra, appeared to con-
verge toward the point of contact, but did not
quite reach it, see fig. 2. This was seen by myself
and the other two observers, and was of great
assistance to us in judging of the time of internal
contacts of the dark body of Venus, with the
Sun's limb. Fig. The 5th, is a representation of the
appearance of Venus at the middle of the egress
and ingress, for the very same phenomenon was
observed at both; at the total ingress, the thread
of light made its appearance with an uncertainty
of several seconds; I judged that the penumbra
was in contact with the Sun's limb 10″ sooner
than the time set down above; in like manner at
the egress the thread of light was wholly broke by
the penumbra. At the egress I found it difficult to
distinguish Venus's limb from the penumbra;
which of course made the second external
contact a little doubtful, and the precise time that
the penumbra left the Sun could not be observed
to any great degree of certainty, at least by me.
Some of the other gentlemen, who were sent to
observe at different places, saw at the ingress and
egress the same phenomenon as we did; though
much less distinct, which was no doubt due to
their telescopes being of less magnifying power
for the penumbra was visible through my
Telescope for the whole Transit, and Dr Solander,
whose telescope magnified more than ours, saw
it, I have reason to think distincter than either Mr
Green or myself; though we both of us saw
enough to convince ourselves, that such a phenomenon did indisputably exist, and we had a
good opportunity to observe it, for every wished
for favourable circumstance attended the whole
of the day, excepting the heat, which was intoler-
able; the thermometer, which hung by the clock
and was exposed to the sun as we were, was one
time as high as 119°. The breadth of the penumbra appeared to me, to be nearly equal to 1/8th of
Venus's semidiameter.

The purely astronomical part of Cook's voyage was
over, though various other observations were made
before the Endeavour left Tahiti on 13 July.

What followed before the expedition returned to
England is well known, but there was one tragedy
which clouded the astronomical work: during the

Figure 7.6. The Cook Monument at Fort Venus.

return, when the Endeavour was near Java, Charles Green died. Not unnaturally, his papers were left in disorder, as he had had no chance to complete his analyses – but the results were there, and the work had been well done.

Most of the instruments used on the voyage still exist, and some were displayed at the "Cook tent" in the grounds of the Old Royal Observatory in Greenwich Park during the tercentenary celebrations of 1975. On Tahiti itself, a pillar marks the site of the old Fort Venus from which Cook carried out his work. In the story of astronomy, as well as that of exploration, the voyage of the Endeavour will always be remembered, and it is sobering to reflect that the name of James Cook would never have become world-famous had he not been sent out to witness the passage of the dark disk of Venus against the brilliant face of the Sun.

Yet despite all the care which had been taken, the results were still not fully consistent. What was needed was a measure of the Sun's parallax, and the values obtained ranged from 8.3 seconds of arc (Planman) to 8.80 seconds of arc (Pingré). This meant a distance range of between 97 000 000 miles and only 92 885 000 miles. In 1824 Johann Encke undertook a rigorous analysis of the whole series of observations, and found that the 1769 transit gave a mean for the parallax determinations of 8.6040 seconds of arc, corresponding to an Earth–Sun distance of 95 045 000 miles (in round numbers). Combining this with the results of the 1761 transit yielded 8.571 seconds of arc, which gives 95 370 000 miles. The modern value is 8.794 seconds of arc; 92 957 000 miles, so that Encke's value was considerably too high.

There the matter rested for more than a century. The next chance would come in 1874.

Chapter 8

Venus: The Transit of 1874

Nobody pretended that the results obtained from the 18th century transits of Venus were satisfactory. There was too much spread in the observations. Encke's derived value for the distance of the Sun – 95 370 000 miles – was generally believed to be too great (as indeed it was, but this was not Encke's fault; he had done his best with the material available to him). Other methods, not involving Venus, indicated that the real distance was more like 93 000 000 miles, but considerable uncertainty remained, and much was expected from the transits of 1874 and 1882. After all, telescopes and astronomical techniques had been improved beyond all recognition since 1769.

Therefore it is not surprising that a huge observational programme was planned. It was truly international; many nations took part, and expeditions were sent out to various parts of the world, some of them to decidedly inaccessible places.

For example, in March 1871 the United States Congress guaranteed a sum of $2000 to equip and send eight parties, five to sites in the southern hemisphere and three in the north. The southern stations were Kerguelen Island in the Indian Ocean, Hobart and Campbelltown in Tasmania, Queenstown in New Zealand and Chatham Island in the Pacific; in the north Nagasaki in Japan, Vladivostok in Siberia and Peking in China were selected.

By now photography had come to the fore, and was expected to play a vital rôle. Each American party was provided with a 5-inch refractor with plate attachment, a transit instrument, and a very accurate pendulum clock. The plan was to photograph the image of Venus as it crossed the Sun rather than concentrate solely upon timing ingress and egress.

There seemed no reason why these new techniques should not work, and in some ways they did, but most of the United States parties had an altogether disproportionate amount of bad luck.

Conditions in Chatham Island and in Tasmania were so poor that little could be done.

Kerguelen fared rather better, but the best results came from Queenstown in the south island of New Zealand, even though the chronometers had to be reset after the journey and the timekeeping may have been affected (a road-side plaque now marks the site of the temporary "observatory").

The Peking party did manage to make some measurements, and there was one additional discovery, not connected with the Sun. The party was led by James Watson, who had discovered several asteroids or minor planets, and on 10 October, using the 5-inch telescope provided for the transit, he found another – No. 139. He invited Prince Kung, the Chinese regent, to give it a name. The Prince chose "Shui Hua Hsing", now used in its shortened form of Juewa.

At Vladivostok, the party led by Asaph Hall – later to achieve fame as the discoverer of the two tiny satellites of Mars, Phobos and Deimos – suffered from unsteady atmospheric conditions, and the photographs were of dubious quality.

Things were better in Nagasaki, where the American expedition led by George Davidson was joined by Captain Yanagi, Chief of the Japanese Hydrographic Bureau, and Hiroma Ueno, a leading astronomical photographer. Fifty plates were taken during the transit. These, together with 47 from Queenstown and fifteen from Kerguelen, were the only really useful pictures obtained by the American expeditions.

There were also more mundane problems. The analyses of the results was held up because the money promised by congress was diverted elsewhere, and eventually one of the chief mathematicians, Simon Newcomb, became so disillusioned that he withdrew from the project altogether. In 1881 David Todd, of the United States Naval Observatory did undertake an analysis and came up with a parallax of 8883 seconds of arc, which did at least bring the estimated length of the astronomical unit down below 92 000 000 miles, but not many people had much faith in it.

Other nations taking part included France, Germany, Holland, Denmark, Austria, Italy and Mexico, but again there was unacceptable scatter (one French expedition, under Jules Janssen, was on its way to Vladivostok when it was caught in a typhoon in the China Sea, and nearly met with disaster). The French results even yielded a parallax of around 9 seconds of arc, which would mean an Earth-Sun distance of between 90 000 000 and 91 000 000 miles – as obviously too small as Encke's estimate had been too great.

Britain made an early decision to send out as many well-equipped parties as possible; the Astronomer Royal, Sir George Airy, was enthusiastic. So was an Army officer Lieutenant-Colonel Tupman, whose part in the proceedings was of special value because the help of the Armed Forces had to be enlisted to take the transit parties to their chosen sites. Colonel Tupman summed matters up succinctly: "No useful measure was neglected. Each observer went out ticketed with this 'personal equation', his senses drilled into a species of martial discipline, his powers absorbed as far as possible in the action of a cosmopolitan observing machine. Instrumental uniformity, and also uniformity of methods, were obtainable and were attained but diversity of judgement unhappily survived the best-directed efforts for their extirpation." Human nature cannot so easily be regimented…

Parties went to Egypt, India and elsewhere, as well as to New Zealand (where the cloud cover was complete). But perhaps the most interesting expedition was that to Kerguelen Island, led by a distinguished Jesuit astronomer, Father S.J. Perry.

Kerguelen is not exactly a holiday resort. It lies in the South Indian Ocean at latitude 49°20′ S, longitude 70°20′ E, roughly midway between Antarctica, Australia and Africa, and close to what is termed the Antarctic Convergence, where cold water from the Antarctic Ocean mixes with the warmer waters from the Indian Ocean to produce a great deal of mist and damp. Kerguelen is chilly and windy, and the surrounding seas are often very rough; on average there are fewer than 130 clear days in every year. There is little vegetation.

Kerguelen was (and still is) a French possession. It was discovered in 1772 by the explorer Yves de Kerguelen-Trémarec, who described it as "as barren as Iceland, and even more uninhabitable and uninhabited". Captain Cook went there during his third journey round the world, and on Christmas Day 1776 the Resolution and the Discovery entered the main bay. Cooke wrote in his diary: "I could very properly have called the island Desolation Island to signalise its sterility, but in order not to deprive M de Kerguelen of the glory of having discovered it, I have called it Kerguelen Land." It is volcanic, and there is one active volcano. Until modern times it was uninhabited, and was left to the seals and the penguins; in 1950 a permanent scientific base was set up, and the summer population can exceed one hundred, but certainly it is one of the remotest parts of the world.

Perry's party left England in May 1874; there were ten ships, the Volage and the Supply. One valued member was Lieutenant Corbet, of the Royal Navy, who was also a Fellow of the Royal Astronomical Society and who was a key member of the team, even though he was only 24 years old. His first view of Kerguelen was not encouraging. "Snow to the water's edge, high mountains in the background... the Straits of Magellan seem to me a sort of paradise comparatively." On landing Corbet and another Naval officer, Lieutenant Coke, set up a makeshift observatory and named it Black Drop Hut; Perry went to a separate site, nicknamed Thunder Bay, and a small station was also set up at "Thumb Peak", under another Naval officer, Lieutenant Goodridge.

Fortunately Cyril Corbet kept a diary, so that we know what conditions were like as the time of the transit drew near. On 7 December: "We got a few observations in the evening, but it was terrible work in the high winds – lamps flickering and blowing out, couldn't hear the ticks of the clocks or anything, and I'm afraid on working them out they will not be very correct." On 7 December: "Trying to keep calm and collected for the day after tomorrow. The barometer is falling fast though, and the weather apparently getting worse." 8 December: "Weather still bad and the barometer very low and still falling, but I shall keep hoping, hoping, hoping for tomorrow. Oh! to think it is so close – I feel funnier today than ever I have felt in my life, and I suppose really tomorrow morning will be about the most unpleasant time of my life up to 11 o'clock, when one will know one's fate... I sat up till one o'clock, observing the stars for a time, and then I lay down dressed on my bed. 9 December: "I did not go to sleep of course, and at 4.30 I was up... Hot cocoa and then to fix my telescope firmly outside the Observatory door, compare clocks and chronometers, and get perfectly ready, although at 6 am it looked dubious, very. I then roused the others, got their fat heads shaken out of them, and we all sat down at our posts patiently to wait; the sky looked very watery, with a thin layer of filmy white or mackerel cloud all over it, with occasional heavy, dark and lower clouds – no wind, everything still and calm."

Corbet's official report then goes on:

11h 49m 43.5s (time by Sidereal Clock) – First appearance of Venus on the Sun's disk: I do not think I could have lost as much as a minute between the above time and true external contact; could detect the break in the Sun's limb almost immediately – it was so steady, and just in the place I was intently watching for it.

12h 1m 25s; Venus half-way on; this was of course only a very rough guess of the middle of Venus.

12h 14m 14.8s. Venus more than three-quarters of the way on at the time; I was its last limb outside the Sun, a very faint light ring distinctly marking it out; the Sun was evidently illuminating that side.

12h 18m 56.1s. I could still distinctly see Venus' second, or last, limb, it being very faintly illuminated; at the above time it very nearly coincided with the Sun's limb; no bluntness in the cusps as yet.

12h 19m 26.8s. This I believe to have been the moment of true contact; the cusps were just beginning to shake and get blunt, and I could distinctly see Venus' second limb, with its faint light in perfect contact with the Sun's limb.

12h 19m 51.0s. At this moment the Black Drop, if any, for there was very little, was at its biggest or greatest stretch, and about to break; but I could all along, since the last time taken, distinctly see Venus' limb inside the Sun's, and only a slight shaky black shade between the two.

1h 20m 8.0s. At this moment there was not more shade between the two limbs, it had disappeared.

Venus never assumed a pear-shape as observed by me, and her loss of roundness was very little if at all; I could from first to last of contact distinctly follow Venus' limb, by its slightly illuminated streak, which pointed it out. The magnifying power I used was 145.

During the whole time I never saw any of the colours of the spectrum – nothing but the blue vault of the heavens, and the white Sun slightly tinted blue by the shade used... I never saw any brightness of wavyness in the first limb of Venus while it was on the Sun; it was the second limb, outside the Sun, that I first was distinctly by the streak of illumination – not at all wavy or uncertain, but quite distinct, and it remained so until all shade had disappeared between it and the Sun's limb, when it disappeared also. The final light streak on Venus' limb was not a bright, glaring or shaky light, but a cold brownish white light, as steady as possible, with black speckles here and there, like irregularities on the planet's surface.

15h 47m 52.0s. From half an hour after internal contact at Ingress till the above time on my Sidereal clock heavy clouds, dark and slow-moving, obscured the Sun; at the time however the Sun was visible for 8 or 10 seconds, and I observed Venus, as near as I could estimate, one-fifth or one-sixth of its diameter gone off; the Sun did not appear again. ... Oh! the happy moment, when from 6 am to 6.30 I had been watching intently the bottom of the Sun for an impression, and I saw it – really and truly the happiest moment of my life... Between Ingress and Egress we hoisted all our flags to the ensign staff and dwelling but, had our breakfast of Oxford sausages and a bottle of champagne, and were very happy; the only wonder to us was – who of the others had to it and what had they got. After noon, the excitement being over, I was too fatigued, after last night's sitting up and this forenoon, to do much beyond writing my official journal and this; it turned out most miserable weather – overcast, cloudy, rain and wind from the SW, and I think we were most wonderfully lucky to get Ingress as we did ... I find that Coke and I differ as much as 15 seconds in our observation of contact, but that is not so very much, as we may not have timed the same phenomenon.

Coke was less than a hundred yards away, had been using a $3\frac{1}{4}$-inch telescope, slightly smaller than Corbet's.

It was three days before contact was established with the other sites. At Observatory Bay, Perry and his companion, Sidgreaves, had observed egress but not ingress; at Thumb Peak Goodridge had seen ingress but not egress. Corbet was very satisfied at his timing of ingress, and the fact that he had at least seen Venus leave the solar disk even though he had not been able to fix the moment of last contact.

The party remained on Kerguelen for some weeks, and finally sailed on 27 February. Corbet's diary ends: "We watched the dreary desolate island for ever so long till all the low land had sunk into the sea and we could see the snow mountains only... We were clear of the land by night, and all with light hearts and full of happiness at getting away from Kerguelen at last after five months of it, which sometimes seems an age, and at other times as nothing but a mad whirling gap in one's existence."

But there is a sad aftermath. Cyril Corbet seemed destined for a brilliant career, but little more than a year after the transit he died of fever off the African coast; and was buried in St Helena. As with Jeremiah Horrocks, so long before, his life was cut tragically short.

We have given Corbet's comments in detail because they may well be a useful guide for the transit of 2004. Of course there were many other observations – notably by John Tebbutt in Australia, and by a team in Mauritius financed by Lord Lindsay and led by David Gill – and the reductions took a long time.

Using the British results, Airy derived a parallax of 8′. 754, corresponding to an Earth-Sun distance of 93 375 000 miles; Edward Stone gave 8′. 88 (92 000 000 miles); Tupman 8′. 81 (92 700 000 miles). Despite the greatest care and effort, transits of Venus simply did not seem able to provide data accurate enough to pinpoint the length of the astronomical unit.

Still, there was one more chance: the transit of 1882, the last until beyond the end of the 20th century. Misgivings or no misgivings, astronomers began to prepare for one final attempt.

Chapter 9

Venus: The Transit of 1882

6 December 1882 was an important day. The transit of Venus was the last to be visible until the 21st century, so that for contemporary astronomers it was literally "now or never".

At least the transit would be visible from accessible countries, at least in part, so that there would be no need to go to sites as remote as Kerguelen. Against this, there was a degree of disillusionment, and some nations, notably Russian and Austria, refused to take any part in the observations on the grounds that there were now better ways of measuring the length of the astronomical unit – notably by using the parallax shifts of Mars and some convenient asteroids.

However, the Transit Commission in the United States remained at least mildly enthusiastic, and autho-rised funds for several expeditions. Incidentally, for the first time there was a surprising amount of popular interest. In New York, where the transit began at 9.30 am, local time and went on for six hours, a tele-scope was set up in Broad Street, near the Stock Exchange, and it was said that "the brokers took as much interest in the transit as others. Broad and Wall Streets were filled with bulls and bears, each with a piece of smoked glass in his hand, and when not engaged in scientifically examining the transit amused themselves by blacking each other's noses and faces". Some State schools closed for the day, and the transit was headline news in most of the national newspapers.

Amateur astronomers, as well as professionals, were of course very much on the alert, and some interesting observations were made, notably of the "Black Drop"

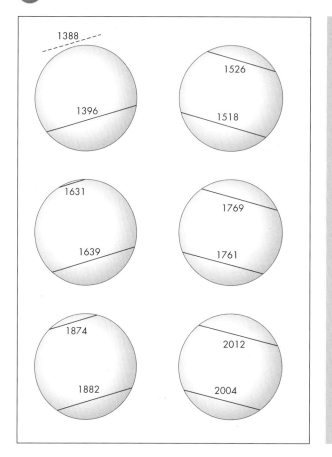

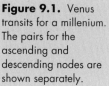

Figure 9.1. Venus transits for a millenium. The pairs for the ascending and descending nodes are shown separately.

and of the luminous ring around Venus due to the atmosphere of the planet. On the whole the weather was favourable.

The main emphasis was on photography, and hundreds of pictures were obtained, all showing Venus as an intensely black disk – much darker than any sunspot. There were also several attempts to see whether there were any trace of a satellite accompanying Venus across the solar disk. There had been various reports of such a satellite, but none had ever been confirmed, and we are now sure that Venus is moonless; the observers had been deceived either by faint stars of by telescopic "ghosts".

Two new observatories were founded specifically to observe the transit. One was in Argentina; the Government purchased some equipment from France, and this led on eventually to the La Plata Observatory, though it does not seem that any useful observations of the 1882 transit were actually made. The second was at Natal, in South Africa.

South Africa was in fact a promising site for the transit, even though the Sun would set before egress. Simon Newcomb led an American party there, and there were also British and German expeditions. As in 1874, nothing was left to chance.

Durban was a favourable site, and this sparked off a long-discussed plan to build an observatory there; the moving spirits were H. Escombe and Professor (later Sir David) Gill. It was decided to make the observatory ready for the transit, and in June 1882 the Durban Corporation voted a sum of £350 for the purpose; another £500 was added by the Legislative Council in the following July, while an 8-inch refractor was presented by Escombe and a 3-inch transit instrument was purchased by the Government. A skilful and energetic Director was needed, and a suitable choice seemed to be Edmund Neville Neison Nevill, an Englishman who had become well-known as the author of a detailed book about the surface of the Moon (it was written under the name of Edmund Neison).* Gill sent him an urgent telegram, offering him the post of Government Astronomer in Natal. Nevill accepted, and sailed hastily for Durban, arriving there on 27 November.

By 1 December he was already at the observatory – only to find that things were far from ideal. The machinery for moving the dome round had been carefully covered with a thick layer of paint. The polarising eyepiece, doubtless excellent in quality, suffered from the slight disadvantage that it would fit neither the telescope nor its accessories. The transit instrument was in Cape Town, and could not be shifted in time. Under the circumstances, it is quite remarkable that Nevill was able to make extremely successful observations of the transit. The one thing in his favour was the weather.

Unfortunately this promising start was not maintained. The Observatory was always under-funded, and Nevill had to do almost all the work by himself. Things became progressively more difficult and finally, in 1911, the Observatory was closed; nothing now remains

---

* *The Moon and the Condition and Configuration of its Surface*, 1875 – when Nevill was still in his twenties. The book, the first major selenographical work in English, contained a full description of every named feature, together with a map which remained as good as any for many years.

of it. Nevill returned home, to live in retirement in Eastbourne in Sussex, where he died in 1938.

British parties went to various sites from Queensland to Bermuda; again there was unacceptable scatter, and all in all it has to be said that the 1882 transit added very little to what had been learned in 1874. It may be useful here to give some details of measurements made over the years.

This last is the modern value, corresponding to mean distance of 92 957 000 miles. It is certainly very accurate.

So, in the end, the transit of Venus method of finding the true length of the astronomical unit failed, although it was certainly not without its value; it marked the start of true international expeditions. And much was learned, even though Edmond Halley's original expectations remained unfulfilled.

Table 9.1. Theoretical Earth–Sun figures compared with experimental measurements

Parallax, seconds of arc	Earth–Sun distance in miles (in round numbers)	
8.0	102 200 00	164 470 460
8.5	96 200 000	154 814 660
8.6	95 050 000	152 963 960
8.7	93 950 000	151 193 730
8.8	92 885 000	149 479 830
9.0	91 840 000	147 798 110
9.5	86 000 000	138 399 800

Authority	Date	Method	Parallax
Maskelyne	1770	Transit of 1769	8.723
Encke	1824	Transits of 1874 and 1882	8.57
Hall	1862	Parallax of Mars	8.841
Galle	1875	Parallax of asteroid Flora	8.873
Airy	1880	Transit of 1874	8.712
Gill	1881	Parallax of Mars	8.78
Todd	1881	Transit of 1874	8.883
Houzeau	1884	Transit of 1882	8.907
Gill	1890	Parallax of asteroids	8.802
Newcomb	1890	Transits of 1874 and 1882	8.79
Hinks	1900	Parallax of asteroid Eros	8.806
Spender Jones	1941	Parallax of asteroid Eros	8.790
NASA	1960	Movements of space-craft Pioneer 5	8.797
NASA	1990	Radar to Venus	8.794

Chapter 10
The Story of Vulcan

Of all the principal planets only Mercury and Venus are closer to the Sun than we are, and so can pass in transit across the solar disk. But could there be any other bodies moving in those torrid regions, which might be tracked in the same way?

There are, of course, some asteroids. Most of the asteroids keep strictly to that part of the Solar System between the orbits of Mars and Jupiter, but there are some which

Figure 10.1. Artist's impression of Icarus close to the Sun. (Paul Doherty)

break away from the main swarm, and may pass close to the Earth. Others approach the Sun to within the orbit of Mercury – such as Phæthon, which at perihelion is less than 14 000 000 miles from the Sun's surface, and must be red-hot (though at aphelion it swings out to well over 100 000 000 miles; it must have a remarkably uncomfortable climate). But Phæthon is a mere three miles in diameter, and other asteroids of the same type are of similar size, so that they are much too small to be seen in transit even with our most powerful telescopes.

Comets, then? There are plenty of those; some almost graze the Sun's surface, and others have been watched as they fall into the Sun and meet with a fiery end. But even a large comet, such as Hale-Bopp of 1997, has a nucleus only a few miles across, and the tenous gases in the coma and tail are virtually transparent. In 1910 the American astronomer B. Ellerman tried to observe a transit of Halley's Comet across the Sun, but, predictably, nothing was seen.

However, little more than a century ago it was widely believed that a much larger body existed, moving well within the orbit of Mercury and ranking as a proper planet. It was even given a name: Vulcan, in honour of the blacksmith of the gods. The search for it went on for many years, and certainly involved the phenomena of transits.

Figure 10.2. Swift's drawing or map of Vulcan as he thought he saw it 29 July, 1978.

The story really goes back to the year 1781, when William Herschel, then an unknown amateur astronomer – by profession an organist at the fashionable resort of Bath Spa – was carrying out a "review of the heavens", using a home-made telescope. He chanced upon an object which was certainly not a star, partly because it showed a disk and partly because it shifted against the background from night to night. He mistook it for a comet, but before long it was found to be a planet; the one we now call Uranus. At once Herschel became famous, and developed into perhaps the greatest observer in astronomical history. Uranus is just visible with the naked eye if you know where to look for it, but it is not surprising that nobody identified it before Herschel did so.

Uranus soon started to misbehave. It persistently wandered away from its predicted path, and clearly something was wrong. There was nothing the matter with the mathematical theory, and eventually it was realised that the trouble was due to an unknown planet, further away from the Sun, which was pulling Uranus out of position. Two mathematicians – Urbain Le Verrier in France and John Couch Adams in England – independently worked out where the new planet must be, and in 1846 Johann Galle and Heinrich D'Arrest, working on Le Verrier's calculations, identified the new planet almost exactly where it had been expected. It was named Neptune.

Le Verrier was born in 1811, and soon became known as a brilliant mathematician. In 1846 he was working in France, though in fact the discovery of Neptune was made from Berlin Observatory (the French astronomers had been rather slow to instigate a search, and patience was never Le Verrier's strong point). Neptune proved to be a giant, much the same size as Uranus, but so remote that it remained well below the naked-eye visibility.

Encouraged by this success, Le Verrier turned his attention to Mercury, which also was moving in a way which was not in perfect accord with theory. The discrepancies between theory and observation were slight, but they were quite measurable, and Le Verrier came to the conclusion that a new planet must be responsible, moving well within the orbit of Mercury. Yet how could it be found? Even Mercury is always inconveniently close to the Sun, and the hypothetical new planet would be much closer-in still, so that it would be drowned in the Sun's rays even at maximum elongation (remember that even at maximum elongation Mercury can never be as much as 30 degrees from the Sun). One possibility would be to catch it during a transit. And in 1859 an amateur astronomer named Lescarbault, living in the Orgères district of France, claimed that such a transit had actually been observed on 26 March of that year.

Le Verrier made haste to travel to Orgères to check upon the observation. What he found must have surprised him. Lescarbault was an amateur astronomer in every sense; he was the local doctor, who also doubled as a carpenter, and his equipment was, to put it mildly, rudimentary. The interview was undoubtedly interesting, in view of the fact that Le Verrier had the reputation of being one of the rudest men who has ever lived (a colleague once said of him that he might not be the most detestable man in France, but was certainly the most detested). The following account comes from the *North British Review* for August 1860: [*]

> On calling at the residence of the modest and unobtrusive medical practitioner, he (Le Verrier) refused to say who he was, but in the most abrupt manner, and in the most authoritative tone, began: "It is then you, Sir, who pretend to have observed the intra-Mercurial planet, and now have committed the grave offence of keeping your observation secret for nine months. I warn you that I have come here with the intention of doing justice to your pretensions, and of demonstrating either that you have been dishonest or deceived. Tell me then, unequivocally, what you have seen." The doctor then explained what he had witnessed, and entered into all the particulars regarding his discovery. On speaking of the rough method adopted to ascertain the period of the first contact, the astronomer inquired what chronometer he had been guided by, and was naturally enough somewhat surprised when the physician pulled out a huge old watch with only minute hands. It had been his faithful companion in his professional journeys, he said; but that would hardly be considered satisfactory qualification for performing so delicate an experiment. The consequence was, that Le Verrier, evidently now beginning to conclude that the whole affair was an imposition or a delusion, exclaimed with some

---

[*] Vol. 33, pp. 1–20.

warmth, "What, that old watch, showing only minutes, how dare you talk of estimating seconds? My suspicions are already too well founded." To this Lescarbault replied, that he had a pendulum by which he counted seconds. This was produced, and found to consist of an ivory ball attached to a silken thread, which, being hung on a nail in the wall, is made to oscillate, and is shown by the watch to beat very nearly seconds. Le Verrier is now puzzled to know how the number of seconds is ascertained, as there is nothing to mark them; but Lescarbault states that with him there is no difficulty whatever in this, as he is accustomed "to feel pulses and count their pulsations", and can at ease carry out the same principle with the pendulum. The telescope is next inspected, and pronounced satisfactory. The astronomer then asks for the original memorandum, which, after some searching, is found "covered with grease and laudanum". There is a mistake of four minutes on it when compared with the doctor's letter, detecting which, the savant declared that the observation has been falsified. An error in the watch regulated by sidereal time accounts for this. Le Verrier now wishes to know how the doctor managed to regulate his watch by sidereal time, and is shown the small telescope by which it is accomplished. Other questions are asked, to be satisfactorily answered. The doctor's rough drafts are attempts to ascertain the distance of the planet from the Sun "from the period of four hours which it required to describe an entire diameter" of that luminary are produced, chalked on a board. Lescarbault's method, he being short of paper, was to take his calculations on a plank, and to make way for fresh ones by planing them off. Not being a mathematician, it may be remarked he had not succeeded in ascertaining the distance of the planet from the Sun.

The end of it all was, that Le Verrier became perfectly satisfied that an intra-Mercurial planet had really been discovered. He congratulated the medical practitioner upon his discovery, and left with the intention of making the facts thus obtained the subject of fresh calculations.

In retrospect if does seem strange that Le Verrier was convinced, but he was confident enough to compute an orbit on the basis of Lescarbault's observation, and he also gave the planet its name. Vulcan's distance from the Sun was said to be 13 082 000 miles, with an orbital period of 19 days 17 hours. The maximum elongation from the Sun was given as 8 degrees, making the planet virtually impossible to see under normal circumstances.

Doubts began to creep in almost at once.

E. Liais, a French astronomer living in Brazil, had been observing the Sun at the same time as Lescarbault, and had seen nothing at all, so that what the good doctor had seen – if anything, was, it was suggested, merely a sunspot. Then, on 20 March 1862, Lummis at Manchester using a $2\frac{3}{4}$-inch refractor, reported a small, perfectly circular spot which moved slowly across the Sun. Two French mathematicians (Valz and Radau) recalculated the orbit of Vulcan, using Lummis' observation as well as Lescarbault's, and came up with very similar results: a distance of between 12 076 000 miles and 13 174 000 miles, and a period between 17d 13h and 19d 22h. Yet confirmation was lacking, and it was only too easy to be misled. For instance, on 30 November 1880 an experienced observer, M. Riccò at Palermo, was observing the Sun with a $3\frac{1}{2}$-inch telescope when he saw several slowly-moving black spots. Closer inspection showed that they were flapping! What Riccò had seen was a distant flock of cranes…

Le Verrier calculated that transits would occur on 22 March 1877 and 15 October 1882. Nothing was seen. Le Verrier died in 1877, but his faith in the reality of Vulcan never wavered.

There was one other possibility of tracking Vulcan – to take advantage of a total solar eclipse, when the Moon covers the Sun completely and the sky darkens (as we have noted, a solar eclipse should really be called a transit of the Moon). At the eclipse of 29 July 1878 two American astronomers, J.C. Watson and Lewis Swift, made independent searches for Vulcan. The fact that the eclipsed Sun lay in the constellation of Cancer, a comparatively barren region of the sky, ought to have made things easier, but the results were not at all convincing. If the observations were reliable, then between them Watson and Swift had detected not one Vulcan, but four – which seemed improbable. Searches during later total eclipses were equally abortive.

The problem was cleared up early in the 20th century, by Einstein's theory of relativity. When this was taken into account, the irregularities in the movements of

Mercury were explained. There was no need for Vulcan, which does not exist – and never did.

There the story ends. Vulcan has passed into astronomical history; it is no more than a ghost in the Solar System.

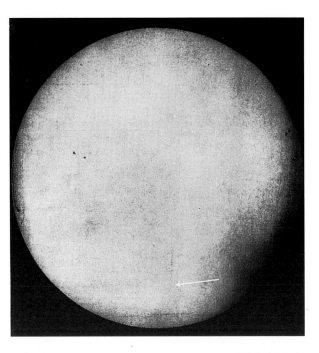

Figure 10.3. The Mercury transit of 1914, November 7.

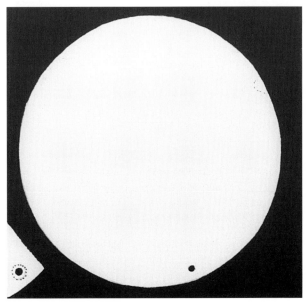

Figure 10.4. The 1953 Mercury transit seen by H.P. Wilkins on November 14.

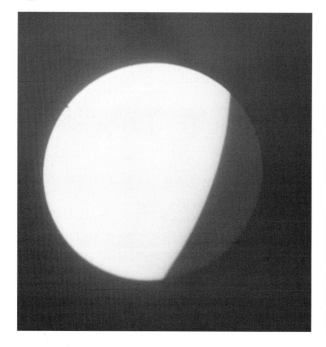

Figure 10.5. A recent Mercury transit in 1970, seen by Terry Moseley on May 9.

Chapter 11
Other Kinds of Transits

So far we have discussed only transits of bodies across the face of the Sun. But there are other kinds of transits also. Most of the planets have satellites, and those of

Figure 11.1. The shadow of Phobos in transit across Mars, a picture taken by Viking from a distance over 13 000 kilometres (8300 miles).

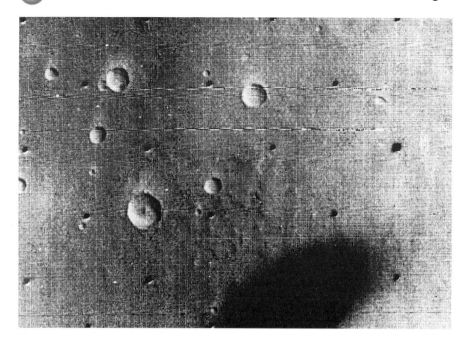

Figure 11.2. A
mini-eclipse on Mars
due to Phobos.

Jupiter and Saturn, at least, are easily visible with very
modest equipment (*really* keen-sighted people can even
glimpse some of the Jovian satellites with the naked eye,
and powerful binoculars show them easily). There is a
great deal of scope here, and the well-equipped amateur
can even carry out some useful work.

Transits of Satellites of the Major Planets

Most of the major transits are given in astronomical
periodicals such as *Sky and Telescope* and *Modern
Astronomer*, and these are reasonably precise, but
sometimes the timings are only approximate and
observers who make accurate measurements can help
in refining future predictions. This can even be done
"at home", because professional observatories simply
do not have the time to carry out all these calculations,
and the home computer can do so. (Here I use the
word "computer" to mean a person, not an electronic
device!). It is also true that some published predictions
do actually rely upon the observational results sent in
by amateurs.

Consider what is known as the "three body problem".

The orbits of two bodies rotating round each other (strictly speaking, round their common centre of gravity) can be worked out to an amazing degree of accuracy, but the ability to deal with three bodies in orbit is so much more difficult that only approximations are possible. These are certainly good in practice, but are little more than estimates based on patient observation – and this is where the amateur comes in to the story.

All the major planets have more than two satellites, making the predications even more complex. When the interactions between the satellites themselves are also taken into account, the complexities are daunting. The satellites can build up harmonic oscillations which force them into orbits with special ratios to each other – as is actually seen in some cases, while in other cases the harmonics remain unstable. Predictions must be based on theoretical models, and these must take into account the equatorial bulges of the planets. The result is a slight doubt about the absolutely precise positions of the satellites, and these affect the times of transits. True, the average over a long period of time will be known well enough, but it is not equally well known how slight the variations and oscillations around that average will be. Only by patient observations can the theoretical models be refined. Many new theories and improvements wait in the wings, and this is how science progresses.

With modern telescopes and electronics, some of the more subtle oscillations around the mean predicted timings will begin to show up. Timings can be refined from hours to minutes to seconds, and so on, as technology progresses. Accurate timings with radio-controlled and other clocks have been routine for many years and modern electronic detectors and associated computers take all this in their stride. Slight variations from predictions can be regarded as a first step toward a much more thorough investigation of the discrepancies. There may even be unpredicted satellite transits, occultations and eclipses, simply because they have not been checked thoroughly enough.

The four "Galilean" satellites of Jupiter are easy to follow. Generally, Io and Europa are easy to see in transit; the less reflective Ganymede and Callisto not so prominently. Shadow transits are easy and the shadows are jet black. With Saturn, Titan is the only satellite whose phenomena can be easily followed, though the really well equipped observer may also find it worth looking out for transits of, at least, Rhea, Tethys and Dione. Racking back through the literature to find out if the object or shadow has been seen before is always fascinating, and may well lead on to some really valuable research.

Other Transits of the Sun

Let us turn next to the possibilities of transits of other bodies across the face of the Sun.

Though Mercury and Venus are the only two large bodies to orbit the Sun at distances less than that of the Earth, many smaller objects invade these regions of the Solar System. The search for Vulcan has long since been abandoned, and Vulcan relegated to the status of a ghost, but there are plenty of comets which can in theory transit the Sun as seen from Earth.

Indeed, at the return of 1910 an unsuccessful attempt was made to observe the transit of Halley's comet. The problem here is, of course, that we now know that the average comet has a nucleus only a few kilometres in diameter, and will be undetectable. The nucleus of the latest really bright comet, Hale-Bopp, was no more than 25 miles across – and by cometary standards, Hale-Bopp ranks as a giant.

Asteroids, then? Those of the Aten and Apollo classes move inside the Earth's orbit, and do undoubtedly transit the Sun sometimes, but again they are in general so small that their transits are bound to be excessively difficult to observe. To be realistic, the difficulties are so great that to undertake long-term searches would be rather a waste of time.

Turn next to transits as seen from other vantage points in the Solar System. To an observer on Mars, the Earth would be an inferior planet, and would therefore pass in transit, as indeed it did on 11 May 1986. Its apparent diameter from Mars would then have been 46.8 seconds of arc. This is much smaller than the disk of Venus seen in transit from the Earth (65.2 seconds) but larger than Mercury (12.9 seconds), so that to our Martian observer the Earth could probably have been seen with the naked eye. Transits of Mercury and Venus also occur from Mars, but would be much less easy to observe.

From Jupiter, the Earth will transit the Sun on 1 January 2002; from Saturn, on 13 January 2005; and from Neptune on six occasions between 2000 and 2005 – but any observers on these remote worlds would find it hard to detect our tiny Earth from such a distance.

However, there will be other opportunities from Mars, which has two tiny satellites, Phobos and Deimos. Both are irregular in shape. Phobos measures $16.8 \times 13.7 \times 11.2$ miles, Deimos only $9.3 \times 7.5 \times 6.2$ miles, so that neither would be of much use in lighting up the dark Martian nights. From Mars, the mean apparent diameter of the Sun is 21 minutes of arc, so that to a Martian observer both the satellites would appear smaller than this; the maximum apparent diameter of Phobos would be 12.3 minutes, of Deimos a mere 2 minutes. Phobos would transit the Sun 1300 times in each Martian year, taking 19 seconds of time to cross the disk; Deimos would on average transit 130 times, taking 1 minute 45 seconds to cross. To our Martian observer, eclipses of the satellites by the shadow of Mars itself would be very frequent, but Phobos would never rise from latitudes greater than 69 degrees north or south, while for Deimos the limiting Martian latitude would be 82 degrees north or south.

From other planets that situation would be different. Of course it is out of the question to land upon the four giant worlds, because of their lack of solid surfaces, but at least it is interesting to speculate.

Jupiter has four large satellites: Io, Europa, Ganymede and Callisto. From Jupiter, the mean apparent diameter of the Sun is 6 minutes 9 seconds of arc, and all four of the "Galileans" would be much larger than this, as shown in Table 11.1 on page 90–91. This means that all four could cover the Sun, to produce total eclipses – though the Sun would be so completely hidden that there would be no prospect of seeing the corona or prominences. But in this connection it is worth remembering that what we call a total solar eclipse, as seen from Earth, is not really an eclipse at all; it is an occultation of the Sun by the Moon – or, if you like, a transit of the Moon.

The minor satellites of Jupiter could pass in transit, and Amalthea, discovered by Barnard in 1892, could hide the Sun completely (its apparent diameter, as seen from Jupiter, would be well over 7 minutes of arc). Of the rest, Himalia, with a real diameter of 106 miles, would have an apparent diameter of over 8 seconds of arc as seen from Jupiter, which is considerably larger

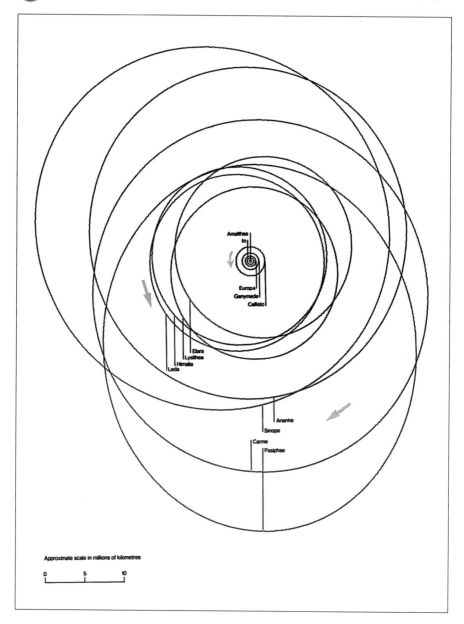

Amalthea
Io
Europa
Ganymede
Callisto

Elara
Lysithea
Himalia
Leda

Ananke
Sinope
Carme
Pasiphae

Approximate scale in millions of kilometres

0   5   10

than either Uranus or Neptune occur from Earth, and transits would be observable. The other minor satellites are much smaller – Leda is no more than 6 miles in diameter, with an apparent diameter as seen from Jupiter of only 0.03 of a second of arc. To detect Leda in transit would indeed require very sensitive equipment.

Saturn is different again; from here the mean diameter of the Sun is 3 minutes, 22 seconds of arc – much the

Figure 11.3. The orbits of Jupiter's satellites.

Figure 11.4. A transit of Io with the Red Spot seen by Voyager.

Figure 11.5. Titan's shadow in transit seen by Patrick Moore through his 10-inch telescope in 1966.

same as the apparent diameter of Uranus as seen from Earth. Of the main satellites, Tethys, Dione, Rhea and Titan could cover the Sun completely. Iapetus, the strange world with one bright and one dark hemisphere moves round the planet at a mean distance of 2.2 million miles, and would have an apparent diameter of 1 minute 48 seconds, so that in transit it could black out a large part of the solar disk – and this does undoubtedly qualify as a transit rather than any sort of eclipse. There would also be observable transits of the small inner satellites.

From Uranus the Sun's apparent diameter is 1 minute 41 seconds of arc; from Neptune, only slightly over 1 minute of arc, so that most of their satellites would produce full "total eclipses". But even if observers could survive there, the chances of observing phenomenon of the satellites would be slim.

Table 11.1. Apparent sizes of the larger satellites from their planet

Object	Angle	
	minutes	seconds
MARS		
Phobos	10	23
Deimos	02	21
JUPITER		
Io	29	51
Europa	16	03
Ganymede	16	56
Callisto	08	47
XVI Metis	01	05
XV Adastrea	00	42
V Amalthea	04	58
XIV Thebe	01	42
SATURN		
VI Titan	14	29
XVIII Pan	02	34
XV Atlas	00	55
Prometheus	03	39
Pandora	02	40
XI Epimetheus	03	08
X Janus	04	24
I Mimas	07	48
II Enceladus	07	24
III Tethys	12	12
V Rhea	09	58
VII Hyperion	00	50
VIII Iapetus	01	23
URANUS		
Miranda	12	47
Ariel	20	55
Umbriel	15	05
Titania	12	27

Figure 11.6. Another Titan transit in 1966, this time seen and drawn by Paul Doherty.

Table 11.1. *Continued*

Object	Angle	
	minutes	seconds
Oberon	08	58
VI Cordelia	01	48
VII Ophelia	01	55
VIII Bianca	02	26
IX Cressida	03	00
X Desdemona	02	58
XI Juliet	03	19
XII Portia	05	37
XIII Rosalind	04	08
XIV Belinda	03	01
XV Puck	06	09
NEPTUNE		
Triton	26	13
Naiad	03	35
Thalassa	06	11
Despina	09	10
Galatea	08	52
Larissa	09	21
Proteus	12	17
PLUTO		
Charon	209	08.85

These angles in Table 11.1 are calculated from the diameters of the satellite and its distance from the primary planet, based on the latest information.

In some cases the angles will differ slightly from the real value because the latest information is often based on different data from that presented here. A word in explanation.

The satellite diameters are direct measurements from the best 'passes' from a space probe, and are likely to be accurate. However, the distance from the primary is often quoted on a different basis. In a few cases it is the actual distance from the surface, giving the correct answer. In some other cases it is the distance from the Planet's centre, a measurement needed for the more fundamental calculations of orbits and gravitational effects, and so on.

In the Table, the apparent angles will, sometimes, be marginally too small, but as this information is intended only as a guide to possible transits, it is accurate enough. The main effect will rarely be of any practical consequence to readers as it will be many years before transits from other planets will be observed. Treat the data as a bit of pipe dreaming or the shape of things to come.

Table 11.2. Apparent size of the Sun from the Planets

Object	Solar diameter		Object	Solar diameter	
	minutes	seconds		minutes	seconds
Mercury	82	39.84	Jupiter	06	09.03
Venus	44	14.47	Saturn	03	21.71
Earth	31	59.95	Uranus	01	40.09
Mars	21	00.10	Neptune	01	03.88
			Pluto	00	48.57

Table 11.2 is a bit of fun as there is always interest in "What if?" questions.

In the Table, the apparent size of the Sun from the major planets is listed. The mean distances of the planets from the Sun are used and the apparent solar size has been calculated to two decimal places. This is part of the fun since that degree of accuracy is quite silly. Only twice in any orbit will the size be anywhere close to the calculated figure, and rarely precisely the same as we have used a "mean" value, not quite the same as an average. Similarly, the orbits are elliptical and the perihelion and aphelion distances will differ by some percentage. In the case of Earth, it is over a percentage point, and about a decade more for Pluto!

The figures are given for the reader to study and decide when there are potential satellite transits from the planetary surfaces when space travel is common. The fun is in deciding when the relative sizes of the satellites switches from being a transit to an eclipse right through to an occultation.

Figure 11.7. Paul Doherty's painting illustrating a future space traveller's view as Umbriel transits the disk of Uranus.

Oddities

On a much lighter note, here are some oddities....

Observing objects in transit across the Sun or Moon is something which is possible almost all the time if you live in the right place. High-flying aircraft are obvious targets, but they do not have to be at a great altitude to be spectacular. One of us (MM) used to work in central London, and one of the things making the train journey home more enjoyable was to watch incoming aircraft on their approach to Heathrow. Some of them might well transit the disk of the setting Sun. It was a truly incongruous sight to see this juxtaposition over the Thames, completely enhanced by the superb orange sky glow from sunlight through the fumes and smoke from Battersea Power Station, which was functional at the time. If the train timings were just right, sometimes the chimneys would occult the solar disk just as a plane

Figure 11.8. Very much a futuristic hope, this artist's impression is the ultimate experience! The space traveller watches from Charon as Pluto transits (strictly occults) the Sun. (Paul Doherty)

Transits down the Ages

transited; at other time the smoke was dense enough for safe observation of the Sun. The Power Station has long been a wreck, and the skies are clearer now, but transit-spotting can still be entertaining. The supreme challenge is to catch Concorde in transit, as I have managed to do twice, though without being able to record it on film!

Lunar transits can also be sought; I have seen Concorde pass across the Moon only once, but it was a sight never to be forgotten.

There are parallels in America. Tours and visits to Mount Wilson Observatory take place regularly, and during the 1970s one of the highlights was to look at the solar disk projected into the basement laboratory of the Solar Tower. Sunspots were much in evidence then, and every two minutes or so a plane's dark shadow transited the side. It seems that it is a matter of honour with the pilots to make sure that their flight-path casts a shadow across the top of the tower. During another visit, for the 1991 solar eclipse, it appeared that the tradition was still being upheld.

The larger Earth-orbiting satellites can also be looked for in transit across the Moon. Remember, however, that attempting observations of this kind with the Sun is fraught with danger unless all the precautions set out in Chapters 13 and 14 of this book are scrupulously followed.

Some of the more esoteric transits occur when high-flying birds are involved. Predictions here are obviously impossible!

Many readers will also remember the spate of UFO sightings in the 1950s. Central Croydon, then the home of one of us (MM) was no exception, and reports came in regularly. Because there was a large school there, which housed an observatory (opened by PM) there were plenty of watchers. At least one explanation turned out to be a large resident owl, which was often seen transiting the Moon or the lit face of the Town Hall clock. The bird roosted and nested in the many trees there. No longer, because the site has been turned into a shopping centre. Such is progress...

Many other UFO reports also involve transits, but not within our definition. The history of astronomy is full of anecdotes of insects and other wildlife crawling across telescope optics.

It is on record that long ago, one earnest amateur using a telescope claimed to have seen large animals on the surface of the Moon. They turned out to be due to ants in his eyepiece.

Chapter 12

Future Transits

Nobody alive has seen Venus in transit and few have witnessed Mercury do it. The long drought is about to break with some of these rare treats almost upon us. By far the best way of getting into the swing of things is to practice on more familiar territory, and Jupiter is just right in the next few years.

Jupiter

The planet is rapidly climbing towards the winter constellations, and is now nicely placed in the autumn skies with a more pleasant temperature in the evening. The 1999 opposition is in October, which means that socially acceptable observing sessions will be possible in that month, and right up to Christmas. The warmer temperatures in early autumn might be the best time to practice techniques for the Mercury transit in November.

During any month dozens of transits and occultations of the four Galilean satellites take place. The transits can be of the satellite itself, or its shadow. Teasing these out from the multitude of spots and markings on the planet's atmosphere is a major undertaking, in the true spirit of observational astronomy. Excellent practice indeed when faced with the small disk of Mercury on the Sun in the middle of Sunspot maximum for the next few years.

A refractor larger than 80 mm (3–4 inches) is the basic starting equipment with the maximum magnification the seeing will allow. Reflectors around

Table 12.1. Some typical transit times for the Galilean satellites

Object	Satellite (hours)	Shadow (hours)
Io	2 ¼	2 ¼
Europa	2 ½	2 ¾
Ganymede	3	3 ¼
Callisto	1 ¾	1 ¾

200 mm (8-inch) have ample specification and have the advantage of (just) being portable. The standard equipment used by one author (PM) is in a fully equipped permanent observatory, and is his 10-inch OG with around 300x magnification. A typical observation of Ganymede itself in transit, and Callisto's shadow is seen in the illustration (see Fig. 16.3, page 143).

Some typical transit times are given in Table 12.1, from which it will be obvious that the passage is a somewhat leisurely process. Spotting the exact times of ingress or egress is the tricky bit, beset with the vagaries of the weather and local seeing.

These are typical times and can be slightly longer or shorter. The precise duration and the ingress and egress times can be found in publications like *Sky & Telescope* for the month or so ahead, or in the *Handbook of the British Astronomical Association* for the whole year. There are computer programmes if you like to calculate the data for much longer periods. We suggest that the monthly predictions are sufficiently daunting to keep any observer fully occupied. Each satellite has an average of two or more transits or shadow transits per month, with sometimes double that or more of occultations of the disk or shadow.

Much of the fascination with Galilean transits lies in the way the orbits are linked together. The three innermost satellites can be involved in simultaneous eclipses and transits, but never in the same phenomena at the same time.

A good way of identifying the transits is by their motion which is always from east to west across the planet. The motion is also a good double check on opposition as the Sun seems to move across the sky and passes behind the Earth. Before opposition a shadow transit must occur before the satellite itself is seen on the planet's disk. After opposition the reverse is true. The real challenge is to find out for yourself what happens on opposition day!

Because the satellites are of an appreciable diameter in a modest telescope, and the transit times are measured in hours, the precise moment predicted in almanacs has to be based on a mean, which invariably means the disk's centre. Observing the exact moment of contact or egress is one of the most difficult of observing tasks, particularly at low altitudes when there is shimmering through a lot of murk. Always plan to start observing some minutes before the prediction, and never give up until at least a quarter of an hour later.

The problem of determining the exact moment is the basic reason why transit observations of Venus never really worked to calculate the size of the solar system. The "Black Drop" effect due to the atmosphere is duplicated nicely on Jupiter. Before dismissing the Venus transit observing skills of the earlier astronomers, a few tries with Jupiter are a salutary lesson of the tests to come with Venus in the next few years.

Saturn

Observing this planet has never been easy and with the ring system opening up, transit phenomena will become rarer in the next few years. Data for Tethys, Dione and Rhea can be calculated from the predictions with transit times half way between occultations and shadow transits half way between eclipses.

These phenomena are a test of both observing skill and seeing. The best many of us can hope for is a 5 minute accuracy. It is possible to pick up Rhea and Dione as points of light in the background sky with a 100 mm (4-inch) reflector and if this telescope is in superb condition in good seeing, claims for picking up Tethys as well have been substantiated. However, this is pushing one's luck too far in most European climes, and at least double this aperture is advised simply to be sure of sorting out the satellites from the background stars. A larger aperture is needed to be confident of seeing a definitive disk during any transit observation, which tends to put these rarely seen events beyond the reach of many.

This is a great pity, because the planet's disk is usually devoid of the confusing detail familiar on Jupiter.

Observing Saturn for shadow transits is highly recommended for anyone wishing to get their eye in, or to test the latest electronic gadgetry.

Venus

The curious pattern of paired transits eight years apart is covered in Chapter 2. It is worth repeating as we prepare for the next set beginning in 2004.

The small 3.4 degree (3° 23′ 39.8″) inclination of Venus' orbit to the Earth means that the possibility of a transit can only occur when the two orbital planes intersect. The Sun is only half a degree in diameter, a small target.

The orbital periods of the two planets takes 1.6 years to line up with the Sun, the only time a transit can occur. This 1.6 year gap is the synodic period, and five of these is almost exactly eight years. The difference is not much, but the faster orbiting by Venus means that the synodic period is very slightly shorter than eight years and by the next set at 16 years the planet has moved too far (more than a quarter degree) out of the Earth's plane. A transit then becomes impossible. All for the sake of a couple of days difference in synodic period.

This close synchonicity of the synodic periods only permits transits during the months of June or December because of the eccentricity of the Earth's orbit. Venus is in a nearly circular orbit.

The gap from 1882 to 2004 is 121 and a half years, but the earlier one back from 1874 to 1769 is only 105 and a half years, so the repeating pattern is neither of these, rather that from a June to June or December to December sequence. Two hundred and forty-three years is a long time to wait to see a repeating pattern and even the shorter 105 year one is still beyond the capacity of modern medical science.

Table 12.2. Recorded Venus transits and for the next two centuries

June	December
	1631
	1639
1761	
1769	
	1874
	1882
2004	
2012	
	2117
	2125

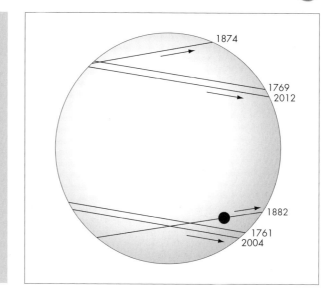

Figure 12.1. The tracks of Venus across the Sun for the four most recent transits and the next two.

That is why the 8th of June in 2004 is so special.

The transit in the southern British Isles will take a full six hours, starting around 05h 15m, finishing around 11h 15m UT. Sunrise on 8 June is around 03h 45m UT, or an hour and a half before ingress. Further north and the Sun rises a lot earlier, the ingress time is later and the duration is somewhat shorter. This is because the planet transits the southern hemisphere of the Sun. An hour and a half after sunrise is not quite an ideal time from the British Isles, although many will say it is not far from it to get the whole session in before lunch. The Sun should have cleared any lingering mistiness and be well away from the horizon at that time after dawn. The whole sequence can be followed at a time of year when the weather is usually settled, that is if the old saying about "Blazing June" is anything to go by!

Perhaps the thing to do is travel east to catch the ingress later in the day where the mid-transit time is close to local noon at 08h 24m UT, something around $3\frac{1}{2}$ hours in time or about 50° longitude E. That suggests north of the Caspian Sea region as anywhere further south into the Middle East to "push" the transit up the solar disk is going to be unbearably hot with the Sun directly overhead. Zenithal observations are notoriously awkward and the British Isles is bad enough towards noon in June. On balance, the British Isles and western Europe in general probably offer as good prospects as any on this occasion. Both authors intend to stay at home for this one. North American observers will miss ingress and those on the west coast will be in a similar quandary to the British Isles eight years later.

There will be no option but to travel east for the next Venus transit on 6 June in 2012. The British Isles will get a tantalising glimpse of the planet just about to egress as the Sun rises. The chances of seeing anything must be vanishingly small, and offer interesting betting odds for the Record Books and insurance companies.

Mid-transit occurs at 01h 36m, placing the most favoured spot at local noon bang in mid-Pacific. Because the transit is on the other side of the ecliptic this time, it takes place across the northern hemisphere of the Sun as seen from the British Isles. Japan and other sites on the east coast of Asia will be very popular for this second, and final, Venus transit in the 21st century. North America will see it well later in their day.

The 2004 and 2012 transits present some interesting challenges to anyone intending to record the event because of the geometry in mid-summer. Dawn is well into the north east octant, and the Sun rises at a sharp angle to the horizon with the North Pole sharply inclined to the left. This axis twists during the six hours of transit which means that the apparent path of the planet across the Sun is not a straight line. It is, of course, but it does require meticulous record keeping when not using a polar drive to follow the Sun (such as a conventional pan and tilt head on a photographic tripod, or an altaz telescope mount), as the position noted at different times can appear most weird.

Anyone making preparations to create a recording of these transits is strongly recommended to invest in a driven equatorial drive which is accurately aligned to the pole. Playback in real time will show the transit as a straight line without a lot of head scratching to get the orientation right. This will be particularly valuable advice when pictures are taken at intervals through a conventional camera.

When viewed at mid-transit in 2004, (as "held" steady by an equatorial drive), the Sun's North Pole will appear to be pointing up at 11 o'clock. Venus will ingress almost at the bottom, ie 6 o'clock, and egress around 3 o'clock, travelling up from left to right across the Southern hemisphere at about 45 degrees as viewed in real life.

From the British Isles at dawn in 2012 the angles are more extreme with the Sun's North Pole pointing almost horizontally left (at about 10 o'clock) with the planet appearing to egress right at the very top of the solar disk at 12 o'clock. All very confusing, although it will be a unique sight in a perfectly clear sky to catch the dark spot on the Sun's disk shortly after a "Green Flash". We can only hope for that combination of miracles.

It is hoped that some intrepid observers will stay home in 2012 to observe such a peculiar event since the visual and recording possibilities are intriguing to say the least.

Much has been said about the the "Black Drop" effect due to what is now known to be the dense atmosphere on Venus, and the bright ring of light around it

Figure 12.2. Altitude adjustments for the 2012 Venus transit. Sunrise is about 03hr 57min UT and transit ends about 04hr 40min UT at Greenwich, 0° longitude. The Sun is then only about 6° altitude. Even with allowance for refraction and other factors, Lands End's conditions are at their limits.

just before and after contacts from the same source (see Fig. 7.3, page 50).

Atmospheric refraction at our horizon can produce some stupendous optical illusions with the Sun's disk appearing to be sliced into stacked layers and other odd shapes, often of unusual colours. The "Green Flash" is just the best known of these effects, with a whole multitude of colourings and distortions being recorded over the years.

Who knows what it will be possible to record with the technology of the second decade of the 21st century? One of the more obvious possibilities is stable flight above the cloud layer to prolong, that is suspend, sunrise indefinitely so that the egress can be seen as though fixed in space. Concorde and similar aircraft have been doing this for years to capture total eclipses of the Sun. The authors look forward to any offers from tour operators willing to try the idea.

Venus' atmosphere has definitely been the "wild card" in transit observations, and as an important aside here, the analogous occultations have also sprung surprises. One author (MM) went to see the occultation of Sigma Sagittari from Kenya in 1981. Through a 110 mm Maksutov (and other small telescopes) the star winked out precisely as predicted and expected and we settled down for reappearance. After a few minutes, there was great excitement at a larger instrument because "The star has not been occulted yet…. no, it's moved round the disk!!". And so it had.

For the whole period when nothing was visible through small instruments, a bright spot of light was clearly visible in a modest aperture moving round the planet's disk to burst forth as full starlight at the exact spot for reappearance seen in the smaller telescopes. At the appointed time.

The moral of this tale is to be prepared for the un-expected with Venus and its atmosphere, and heed the experiences of the past at transit time. Because seeing conditions seem to be of lesser importance than optical quality and resolution, intending observers should carry out tests and trials as early as possible to find the best equipment. The trade-off has to be between large apertures for quality as against the huge heating effect of full sunlight. For the special case of sunrise, the latter is so much less critical, and the largest of tele-scopes can be employed for a change.

Because there have been reports of bright spots seen on the planet's disk, heed the warnings in the chapter on observing. Mirror optics can give ghost images, the most likely cause of these spots. However, because of our experience at the Sigma Sagittari occultation, sheer size and hence optical quality does seem to be the most desirable feature in the observing equipment.

Mercury

Figure 12.3.
Mercury transits for 150 years. The next one in 1999 is almost as bad as it can get.

Whereas Venus' orbit is almost circular, Mercury's is highly eccentric and inclined over 7 degrees, more than double. The immediate benefit is a much greater probability of transits, and that turns out to be true. Instead of a drought of over a century the events can occur at much more regular intervals. When the Earth's own orbital eccentricity is added into the calculations, at the descending node a transit can only occur in May as our planet dips below the mean orbital plane. In November the opposite occurs at the rising node. The combined

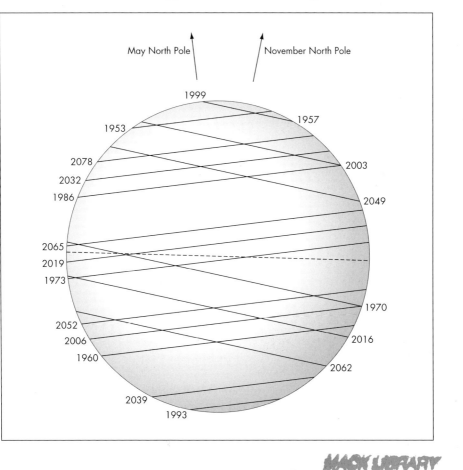

Table 12.3. Mercury transits in the 20th and early 21st centuries

Date	November day	May day	Interval years
1907	14		
1914	7		7
1924		8	7
1927	10		3
1937		11	10
1940	11		13
1953	14		13
1957		6	4
1960	7		3
1970		9	10
1973	10		3
1986	13		13
1993	6		7
1999	15		6
2003		7	4
2006	8		3
2016		9	10
2019	11		3

eccentricities conspire to make the November transits more frequent than the May ones by the 7: 3 ratio mentioned before. This is in marked contrast to Venus where there is the same matched pairing or double transits at each node at regular intervals.

Kepler predicted the first transit to be observed by Gassendi in 1631. Recent events are listed in Table 12.3, from which it can be seen that future ones are much more difficult to unravel than in the case of Venus, particularly if you try to apply the 7:3 ratio. The full picture only emerges after many centuries.

There is said to be a 13, 7, 10 and 3 years sequence between transits. Predictions based on that rule appear to break down in this Table, and it is only given for information.

The 1999 Mercury Transit

Mercury transits make these special events a very useful opportunity to practice for the Venus pair. We have two coming up before 2004, with November 1999 a prime target.

Figure 12.4.
Transit of Mercury, seen from Carlisle, at approximately 12.50 pm on 9 November 1973. It was a cloudy day in Carlisle and this picture was taken in desperation through thin, quickly moving cloud, towards the end of the transit. The photograph was taken with an SLR camera at the prime focus of an 8″ Frank's reflector, which was stopped down to a very small aperture using a cardboard disk. A dark sun filter was also used to further reduce the light and heat. (David Ramshaw)

It has to be said that the science to be gained from a Mercury transit is close to zero. What makes them interesting is this chance to practice and check out the equipment in the field. Then there is the intellectual challenge of one of the more difficult observations one can carry out in the daytime as distinct from conventional night-based astronomy. Regard the chance for a lot of foreign travel as a bonus.

The planet's disk is very small and the sunlight's intensity is as high as ever, placing a premium on optical quality and clear skies. The geometry conspires to make the May transits longer than in November, and these can take up to nine hours. The sheer length of time needed places yet another constraint requiring clear skies for long periods, even with the shorter transits in November. Perhaps the safest bet is in the tropics, and our expedition to see the 1999 event has opted for the Pacific Basin.

In 1999, mid-transit is 21h 41 to 42m UT, which makes Indonesia an ideal region for observation.

Having said that, the 1999 circumstances are almost as bad as they could be because the orbit geometry conspires to make the event an almost near-grazing one.

Travelling only a relatively short way to New Zealand will guarantee not seeing any of the planet's disk fully in transit. Observers at the very northern tip of the North Island might, just, get the best of this truly grazing phenomenon. It will require nerves of steel to observe from there to see this rarest of all events. Recent predictions indicate that the whole of New Zealand will miss out completely on egress, interior contact will not happen. If the skies are clear on the day, there will be some striking and unique pictures emerging from New Zealand with several minutes opportunity to capture the "Black Drop" effects and other optical illusions at the solar limb.

The western coast of America will catch something before sunset, whilst the densely populated parts of north-eastern USA will be in early night darkness and miss out completely. For anyone in America, travelling to the greater Pacific Ocean will be a much better bet. Should they decide to stay, the first contact or ingress point will be around 21h 11m with a position angle just over 30 degrees. The fourth contact or egress point occurs under half an hour later, about 26 minutes in fact, and will be seen at a position angle close to 13 degrees.

Right across the Pacific it will be sunrise, another awkward observing session at the best of times in the late autumn. You will need to be at the extreme edge of the continent for any chance of success, which means that Japan will be a prime site with its high concentration of active astronomers. Ingress starts around 21h 15m with similar position angles, duration and egress times as the western USA.

Because these two extremes of the observing window have a high concentration of observers, many will opt to stay at home and some high quality observations will be made from the major professional observatories. Mount Wilson and the McMath on Kitt Peak are obvious centres of activity on the day. Hawaii is not placed in either end of the home territory of the main astronomical communities, yet has one of the best solar observing sites in the "window of opportunity". Mauna Kea will be a Mecca and a popular choice.

We think Indonesia is the better all-round region for the amateur, with a wider range of alternative holiday possibilities an added bonus.

That leaves Australasia as a viable alternative. Early spring can be a delightful time to visit. As the planet will be that much closer to a grazing phenomenon in those places well away from the tropical heat, consider travelling there for the special sights. One thing can be guaranteed in much of this option, you will be very much on your own, in every sense of the word! A true expedition in the true spirit of the Victorian Age, a century or more before.

Only approximate timings and position angles can be quoted in advance because the observing conditions are the deciding factor, particularly if somewhere on the extremes is chosen. That is the whole point of the exercise, to investigate the limits of the human eye and modern equipment. It will be the next best thing to true science to come out of a transit observation.

With enough observers spread about, and there will be a large number in USA, Japan and Australasia staying at home at their normal observatories, observing conditions will be as varied as they can ever be for a single event. This transit provides a superb chance to gain some genuine science from something no longer associated with the serious side of the business.

Modern electronic detectors can be arrayed against the camera and the naked eye. Perhaps we can really sort out the "Black Drop" effect and the limiting conditions when it is likely to occur. Many of the effects definitely arise from optical illusions in the eye/brain complex when viewing with equipment on the borderlines of resolution and contrast. The solar limb always "boils" on a clear day and at dawn and dusk the turbulence is considerable with atmospheric stratification making the edge "jump" in quite unpredictable directions. The limb is much dimmer than the main disk, and a lot redder. All these factors guarantee to limit the precision when a dim (and small in small apertures) spot makes its first appearance.

Then there are all the improvements in telescope technology to add into the equation. Many amateurs regularly use apertures considerably larger than professionals of only a few years back. Even a modest portable amateur set-up is a lot more effective than any taken on expedition in the heroic days of yore.

November 1999 at the Mercury transit will be a major breakthrough in amateur science. It is only by collating umpteen observations under such a diverse set of circumstances and equipment that science can progress. We know that professional observatories are at the leading edge, and will experience fewer technical difficulties. The "spin-off" from them will be less interesting. It is the sheer number and variety of amateur equipment, much of it a lot better than some professional sites, which blurs the distinction these days and makes the whole prospect so exciting.

If you do not have plans for 1999, get ready for 2003.

From the 1999 observations much more precise predictions can be made for that next Mercury transit and from them the more impressive Venus events. Exact ingress, egress times and position angles and duration can be calculated more accurately from the 1999 data in terms of what is feasible in practice. From that it will be possible to recommend specific sites and equipment to confirm or deny those predictions. Where, indeed, the special and rare impressive grazing spectacle is best seen in future.

We cannot over emphasise the excitement of making a contribution to science when participating in any astronomical sight. Do not be put off by the apparent simplicity of any equipment you have. For solar projection, almost anything will do. The whole point is that a negative observation is just as valuable in setting the practical limits. You never know, some unique sky clarity or seeing might just buck the odds, and you could be the only one involved. Never forget the vagaries of the weather.

This happened in 1998 at the annular eclipse of the Sun in Sumatra. We had an almost unique view in good skies in a small "window" of opportunity whilst virtually the whole of the rest of Asia was under that terrible flooding and rain for weeks.

The next Mercury transit opportunity is a little more favourable on 7 May 2003.

Mid-transit is predicted for 7h 53m UT, making the whole of Europe quite favourable. North America is almost a complete dead loss again except for the latter stages. Nevertheless, if some spectacular sunrise effects are needed, very much in the style predicted in Europe for Venus in 2012, much of western USA will then be extremely favourable. Make preparations now, and do the calculations. Refine them in the light of the 1999 experiences.

Three years later in 2006 and two after the Venus transit, another Mercury event occurs with a mid-point at 21h 42m UT. Java as an observing site is a hot favourite again. Western USA should see it around local noon.

9 May in 2016 and 11 November in 2019 both have mid transit times in early afternoon UT. European observers will be favoured for them.

Part 2

Observing Transits

Chapter 13

Your Safety

It's a sad reflection on our times that an author can be held responsible for not warning readers about common sense matters affecting their own well being! That being the case what follows is a distillation of some self-evident things to do and not to do, things that will make observing transits as safe as possible. Transit observations are quite specialised and because they do have an element of real science behind them, many people are taking up the work in the hope of making some fundamental discovery. We'll start at the beginning.

Looking through a telescope at any astronomical object at night is about as safe as anything can be, but there are a few occasions when you could risk injury. The first and most obvious thing to do is to ensure that you know your way around the observing area, and don't leave objects lying about to trip over. All very sensible (whatever your astronomical interest is), if you have a permanent observatory building, but not so easy to achieve when on location or when using a portable instrument like a commercially-made Schmidt-Cassegrain Telescope (SCT) that you can cart about and set up anywhere at a moment's notice. Get to know the equipment and how to assemble it (and then avoid its sharp corners) in total darkness – why are so many SCTs finished in dark enamel, and equipped with black wiring? They look great in the shop, but are invisible in the darkness!

Perhaps the most important thing to do is to acquaint yourself with the way the telescope fits together with yourself. Not such a stupid remark, really, because all too many people will tell you that it is the easiest thing in the world to jam your eye into the eyepiece when you're not too sure where it is in space, in the pitch black.

Sitting glued to the eyepiece for long periods waiting for the crucial timings in transits is par for the course, and it is all too easy to get into a mentally numbed state and forget or lose your orientation.

Spend some time practising in the daylight to get your observing position properly sorted out for maximum comfort in the cold and damp of an observing site. Then check over the eyepiece itself. The safest ones have rubber eye-cups. If the diameter of the eyepiece is at least as big as the bony part of your skull around the eye socket, and padded, it will help protect your eye if the worst happens.

There is another incidental benefit of this simple idea. Because the eyepiece padding will now extend way beyond the pupil size, it will give excellent stray light shielding from the surrounding sky, and vastly improve contrast whilst reducing eyestrain.

The point about eyestrain is important in the relatively longish sessions needed for transit work when the whole point is to discover or check the real timings. All too few telescope eyepieces come fitted with a proper light baffle which can act as a sensible eye protection device at the same time. One of those "floppy" funnel-like things seen on gunsights is ideal. Why? They are designed just for the purpose we need, to protect the gunner from recoil – jamming the eye into the eyepiece at high speed in our excitement in the pitch dark. We believe this is a much-overlooked point in telescope design.

Armchair astronomers – those that observe remotely – need not feel too smug if they have read this far. Computer monitors and TV screens for electronic visualisation are not without their problems.

"Screen radiation" is much publicised and is largely hype. Colour cathode ray tube screens do have a radioactivity above the general background (sufficient to classify most as low-level radioactive waste and thereby invoke protest were they to come out of a power station, in fact). They are not so classified because the screen ingredients contain materials with *natural* radioactivity, like potassium. Paradoxically, we are dealing here with natural radioactivity which is in a man-made edifice!

The other reason why it is not a good idea to get your face too close to a screen is that low-level x-rays are emitted when the screen is on. The larger the screen the higher the level, and with the trend towards bigger and bigger monitors these days, the problem is not going to go away until semiconductor-technology "flat screen monitors" are cheap enough to replace the cathode ray tubes. Filters and meshes sold to protect against "screen radiation" are next to useless in stopping these x-rays from the larger monitor screens.

Reading glasses actually make a much better x-ray filter, and will stop you squinting too close as well. Keep your head at least a normal reading distance, say around a foot (30 cm) away from the screen.

X-ray emission isn't really a safety hazard under normal circumstances, of course. The under-discussed factors of flicker rate and light intensity levels are more important. Forget the technicalities here. The important things to sort out are a nice steady screen, not too bright. Modern monitors with the "VGA" and "SVGA" compatibility, used with the right driver software, are usually excellent. If there is a choice, opt for the highest screen-refresh rate.

This is particularly important for anyone who suffers from epilepsy; if you do, you will almost certainly know about the possible effects of flashing light. If you don't, ask your doctor.

The old "green-on-black" monitors weren't good from the point of eye-strain. The eye is most sensitive in this green region in the spectrum but because the bulk of the screen is black the pupil dilates automatically. The brain thinks it is night-time, yet is getting what is in fact an extremely intense burst of green light. Furthermore, the extremely small image from the lettering is insufficient to activate the iris stopping down system, which overloads the sensing mechanism, and hence the brain. Most people seem to step up the screen brightness/intensity to make the lettering stand out better, making the situation even worse. Hardly surprising then that the person quickly gets headaches and fatigue as the brain rapidly gives up the uneven struggle with the paradox.

If you're using an old monitor for some mundane task, just keep the brightness low. Or switch to a reversed (dark letters on a light background) image so that the eye sees a wide area of brightness and the pupil contracts correctly, much improving eyesight deficiencies at the same time!

All this might appear to be making a mountain out of a molehill. Not so. To do transit work with any confidence when timings are not known accurately can mean many minutes, even hours of peering at a screen to get the moment accurate to a fraction of a second. If you are fatigued or suffering eyestrain, your personal equation will be way out. Timings are no longer reliable.

What this means in practice is that whether you depend on the old-fashioned "Mark 1 Eyeball" or the latest SVGA monitor screen attached to a state-of-the-art computer system, keeping the eyesight in first rate condition is all-important.

The Moon

Before leaving the question of eyesight, there is a night-time sight which does need a little more care than might at first seem obvious. That is the Moon.

Nearly all transits involve the passage of some celestial object in front of another, and these are all perfectly safe to look at with any optical aid you like. The possible exception is the Moon through a very large telescope. The light intensity under these circumstances is not dangerous in itself but it can be quite uncomfortable to look at. In long observing sessions, headaches and eyesight fatigue set in. A filter of some sort is usually enough to stop the problem. Filters are readily available for all makes of telescope. Sometimes a better option is stop down the main optics, which has an incidental advantage of improving the contrast.

This special case of lunar observing does seem unlikely to be of much practical consequence in transit studies, unless, of course, an asteroid or similar body gets too close to the Earth for comfort. If that happens the literal impact on civilisation would be of more concern!

The much more likely and practical situation is the increasing fascination, almost a fashion, of trying to spot high flying aircraft and other man-made objects transiting the lunar disk. One of the most impressive sights is to see Concorde crossing the lunar disk. Ordinary camera lenses are ideally suited for this job and no special precautions are needed except to be aware of the Sun and where it is in the sky. Take special care with thin crescent Moons near to sunrise and sunset.

The question of eyestrain is obviously most serious at full Moon, and has become more of a problem today with the spate of large aperture and fast focal ratio telescopes in the marketplace. These are perfect for their intended purpose of gathering as much light as possible for deep-sky work, which makes them a little "over-kill" with the bright Lunarscapes. We are not concerned with shortish glimpses here rather longish sessions of intense scrutiny in case an object such as a satellite transits.

The large apertures and fast focal ratios of modern mirror telescopes do not help too much. Much better to use an older type of refractor with a long focal length and a relatively fainter image, or stick with a camera telephoto lens. The added advantage of the latter types of lenses is that many of them can be adapted to fit straight onto a video or charge-coupled device (CCD) and thereby free up the prime mirror telescope for its main purpose, which is faint object astronomy.

The Sun

SPECIAL PRECAUTIONS FOR SOLAR OBSERVATIONS

We now turn to the Sun, and the very special and crucial safety aspects which must be addressed. Planetary transits across the Sun are rare, and as some interesting ones are coming up, now is the time to get ready and be prepared for these special events.

We can also regard some situations at solar eclipses as a version of a transit. Transits of the lunar disk across sunspots are always fascinating, and with sunspot maximum approaching as we write, there will be many opportunities in the next few years. Such transits are a supreme test of photographic skill, particularly if the jagged edge of the Moon is enlarged enough to be seen astride a sunspot.

Imaging high flying aircraft and satellites is a virtually untapped source of fun, as well as being good practice for the serious business of astronomical and planetary transits.

The simple advice for looking directly at the Sun at any time is very blunt. Don't do it, don't risk your eyesight.

There is *no* safe way to look at the Sun with the naked eye. Do not even think of doing it.

Even when there is just just a sliver of the disk left above the horizon, it can be hazardous.

The only time it is perfectly safe to look at the Sun with the naked eye is during the brief moments of a total solar eclipse. That's the whole point of eclipse viewing! Official guidelines are available for safe viewing the Sun at any time and during the partial phases of an eclipse. Read, learn and inwardly digest these guidelines. What follows is a brief interpretation.

Nature has granted us a very simple and effective safety device to protect our eyes, the "blink response". In normal life we never need to look in the Sun's direction, but if we do so by accident, the intense brightness and surprise guarantees that that we blink and look away. This is the "blink response". Stick with it and do not be tempted to over-ride nature and stare at the Sun for any length of time.

There are very good reasons why nature knows best – well, better than you do, anyway.

The eye has no pain sensors or nerves on the light sensitive part at the back of the eyeball – the retina – which means that it is perfectly possible to fry your eyeball without being aware of it. The damage is done by heat and invisible long-wavelength radiation (which is given the technical name of infrared, sometimes abbreviated to IR). Even when the Sun is quite low down on the horizon, sunlight can feel warm. That is a real danger period when temptation is at its highest. Whenever the sunlight has any warmth left in it, never look directly at the Sun. Better to be safe than sorry.

At the other end of the spectrum, there is ultraviolet (sometimes abbreviated to UV). This is energy at a shorter wavelength than we can see, and is bad news as far as the eye is concerned. UV action on the retina is poorly documented, but is certainly damaging to colour vision and acuity in the long term. UV action on the cornea is a major contributor to cataract.

If anyone doubts the need to avoid looking directly at the Sun just try the effect of focusing the Sun's image through a magnifying glass onto paper. We have all done it at some time to set paper alight. Magnifying glasses used to be called 'burning glasses' for just that reason.

Don't even think about looking directly at the Sun through any form of optical device. That includes all types of viewfinder on cameras.

Special solar filters for telescopes and camera lenses are available and these demand even more care and attention to detail. They merit a special section considered in the next chapter.

Chapter 14

Observing the Sun: Full-Aperture Filters

For almost all occasions the only perfectly safe way to view the Sun is to project the image, and that is our next major step, reviewed in the next chapter. Before doing so, there is a risk with some types of telescope which is of such potentially fatal consequences that it merits repetition in some detail here.

Reflecting and Catadioptric Telescopes

Modern telescopes are just too good.

Up to the 1980s virtually all telescopes were in the hands of astronomical enthusiasts who knew what they were doing. Long focal lengths, around f/8, and the Newtonian design was almost universal with a 6-inch diameter mirror regarded as quite a large telescope. Eight-inch reflectors had very proud owners.

These days, 8-inch (200 mm) telescopes working around f/5 or better are regarded as "entry level" for amateur astronomers. Herein lies the grave danger.

These highly efficient telescopes, in the hands of the inexperienced, can be dangerous. The advice you read in some of the older amateur astronomy books no longer applies: for example, you should not even attempt to use a large telescope for solar projection (often recommended) until you have consulted a real expert at a local Astronomical Society first.

Commercially-made Schmidt-Cassegrain telescopes (SCTs), for example, have large apertures (around 10-inch is typical) and short focal ratios. The large apertures are bad enough but the short focal ratios mean that a huge amount of light and particularly heat energy is concentrated into a very small spot or area. If an eyepiece is not fitted, the emergent rays can inflict severe skin burns almost instantly, far faster than you could snatch you hand away.

Nor are eyepieces designed to accept such heat. Even perfectly clean glass will absorb a lot of energy under these conditions, and heat up beyond working tolerances. In a real world no eyepiece is likely to be perfectly clean, so more energy is absorbed. Some eyepieces contain delicate bloomed surfaces, or are made with rare materials not intended to be heated. Result – a cracked or ruined eyepiece. It can be even worse. Some eyepieces have balsams or other mountants, and many will contain traces of moisture. Most eyepiece designs are completely sealed to stop moisture and dust getting in. Whatever the design or condition the end result can be a catastrophic heating of the air or moisture in the inner cavities. It's like pulling the trigger on a firearms cartridge, and with shattered glass as the missile, the end results simply don't bear thinking about.

Don't even think about using a large mirror telescope for solar projection – see Chapter 14.

Viewfinders

Most telescopes, even the cheapest, are generally fitted with some kind of viewfinder, usually a smaller "piggy-back" mounted telescope, often fitted with cross-hairs. *All such finders and any other optical gadgets on the side must be removed or fully blanked off.* Cap the viewfinder *at the front* with something that won't fall off. There is a very real danger of forgetting that you are observing the Sun and absent-mindedly looking through the viewfinder; if you do this, you could be blinded. Less important in the long term, but very annoying, is the fact that the lens of a viewfinder telescope focuses the Sun's ray directly on its cross-hair sight, which may disappear in a puff of (expensive) smoke.

Filters

The only practical alternative to projection, for solar observing, is a filter of some sort.

First and foremost it has to be said that this filter must be one designed for the job and given a guarantee of quality.

A few telescopes (not from reputable manufacturers!) are still sold with what is described as a solar filter which is to be placed behind the eyepiece. In our opinion, sale of such a combination constitutes criminal recklessness, as the potential hazard is all too terrible to contemplate. If it cracks, as is highly likely with just a modest telescope aperture, the eye is exposed to the full terror of concentrated sunlight and heating. Nobody has quick enough reactions to avoid sight loss or the whole eye in these circumstances. Flying glass just adds to the potential nightmare.

The only form of filter to look at is one designed to fit over the main aperture, so that only the filtered light passes into the telescope. These are known as "full-aperture" solar filters.

The quality of the filter must be excellent. Buy only the best, from a 'known' manufacturer. If it seems expensive, regard it in the same way as you regard an insurance premium against sight loss! British Standards, CE and NBS exist for your safety. Items carrying such marks have a quality guarantee.

In a tropical country with crystal clear skies, about 1/100 000 or less of the Sun's normal light is needed for comfortable viewing indefinitely. This light reduction is identified as a neutral density filter, "ND 5", or as a welding glass equivalent of "Shade 14". There are other schemes but these are the ones to go for because they indicate an approved measure of safety from BS, CE or NBS, and should not be too far from the claimed light reduction.

ND 5 is quite often much too dense for a normal northern climate and a frequent complaint is that the Sun cannot be seen at all! Something close to ND 4 (Shade 13 in welding glass) is often a much better bet for the more common slightly opaque conditions.

This means that it is almost impossible to recommend a single filter which is comfortable and safe to use for all conditions. If the budget will stand it, consider buying two or more filters for naked eye viewing through a telescope. We always carry three different density filters, ranging from ND 3 to 5, or the welding glass equivalent of Shades 12 to 14.

It is not necessary to wait for a perfectly clear day to test a filter, and this simple rule of thumb method is a good way to grade them in advance, should they get mixed up out of their protective boxes. If the filament of a 100 watt clear lamp can just be made out as seen through the filter then it's likely to be at least ND4 or Shade 13. If nothing is seen, or just barely, then ND5 or denser is indicated. If the filament can be seen at all brightly, then mark that filter very clearly indeed and keep it for a very cloudy day…

Commercially-Made Solar Filters

Remember, pay extra if necessary, but be sure to buy any solar filter from a reputable trader, ideally manufactured by one of the major instrument makers. You can select from these types.

Full-Aperture Filters – Welding Glass

For long-term durability, this is the best type. It is specifically designed to allow welders to work long shifts in complete eye safety. The glass is rigorously tested and has a test mark and density grading. These can be used with peace of mind under all conditions, including grubby finger marks and scratches, and only become unusable if broken.

Don't consider *any* dark glass which does not carry a proper test mark and grading stamp.

Welding glass comes in a variety of colours and usually imparts a greenish tinge which is quite pleasant. The rough edges when sold "rough cut" can be sharp. If you get hold of a rough-edged piece, simply bind the edges with electrical insulating tape. The colours available quickly identify the shade.

When not in use keep in a polythene bag or bubble pack. Clean with anything used for glassware and these filters will last a lifetime.

ADVANTAGES – Cheap and durable.

DISADVANTAGES – Flatness and evenness of density are a matter of luck. Welding glass is never optically flat. It is extremely difficult to find a cheap source which does not cause doubled images (due to reflections off both sides), or poor focus. Then there is the colour.

Full-Aperture Filters – Mylar

The name really applies to the kind of plastic, not to the filter itself. These solar filters are actually a very thin sheet of Mylar which has been coated with aluminium. The correct stuff to use has the aluminium applied by evaporation in a vacuum (it is used in the food industry for packaging).

Material coated on one side is sold, and perfectly usable if it passes the "100 watt bulb" test, but this is never a good idea in inexperienced hands because that surface can get damaged easily. We do *not* recommend single-sided Mylar for solar filters.

Nearly all the "viewing spectacles" sold for eclipses have either two thicknesses of this single coated stuff (coating inwards), or Mylar coated on both sides.

Mylar tends to be on the dense side, so that it can be used in the brightest of skies (for safety reasons). The image tends to be a bright bluish tinge, which is not a normal Sun colour, but one soon gets used to it.

Provided the sheets are bought for use as solar filters and from a reputable source, ideally with a CE (in Europe) or similar guarantee, they should last for years – although the basic idea is "single or disposable use".

Keep Mylar filters in a bone-dry place and protect them at all times from scratches and any rough handling. Reject any filter which shows uneven density across the width, and definitely throw away any with pinholes of bright patches.

Keeping Mylar filters in a dry place is vitally important, as traces of moisture will eventually dissolve the aluminium. Rapid changes of temperature mean more than just a casual check over in case condensation has formed on the coating. Salt air is particularly destructive to aluminium in such thin layers. Anyone living close to the sea must remember to check Mylar filters regularly.

The simple rules for keeping Mylar going are:

No moisture, no salt air and definitely no touching or fingering. Keep in a sealed polythene bag when not in use and they should last for years, but check them over for damage each and every time you get them out.

ADVANTAGES – cheap.

DISADVANTAGES – Very easily damaged in the size necessary for telescopes. The colour imparted to the image. Polarises the image.

Full-Aperture Filters – Film

It is possible to use overexposed and processed film as a full-aperture filter. There is considerable confusion on this subject, but the safety rules are:

1. Only old-fashioned black and white (mono-chrome) film is safe.
2. Colour film of any sort must never be used under any circumstances.

Almost any brand and type of black and white film can be used. It is also a good way of using up old film. Expose to ordinary daylight for a few seconds, develop in any developer for several times the normal time, then fix and dry normally. Pieces of this mounted in cardboard mounts work perfectly as a "naked eye" solar viewer, and last for years if free from finger marks or scuffing.

Remember, colour films are really dangerous (even when appearing black) as they are transparent to IR. Use monochrome only.

The Sun's image is a normal yellow colour, and the point about these filters is that they can be made to order in an almost inexhaustible supply for virtually no cost if you do your own processing, or the cost of professional processing if you don't – ask for "pro-cessing only, no prints". A wide range of densities can be made to cope with all sky conditions. Check them over with the "100 watt bulb" test and grade them accordingly.

ADVANTAGES – cheap. Can be home-made in inexhaustible supply.

DISADVANTAGES – extremely difficult to predict quality.

Full-Aperture Filters – Metal

Anyone using specialised metal-on-glass filters is unlikely to be a novice, and will know how to use and keep these in condition. *Inconel*, which is stainless steel, chromium and silver are the most common materials in use. Commercially-made inconel filters from a reputable supplier will be safe if free from pinholes. Inconel filters are, however, generally expensive. They are designed and sold for indefinitely long naked eye views, and come into their own for photography.

> ADVANTAGES – the best option: safe, the Sun is not coloured, the best ones are optically flat, they are available in a range of densities, they are generally supplied in mounts designed for the most popular commercially-made telescopes.

> DISADVANTAGES – only cost.

Other Types of Full-Aperture Filter

The advice is simple: don't use anything apart from those listed above. None can be recommended.

Special mention needs to be made about smoked glass. This was the favoured filter until recently, particularly for naked-eye observations, but is now recognised as totally unsafe. Some examples are claimed to be safe when tested in a laboratory. Fair enough – that single filter is (probably) proven, but what about others that are, or seem to be, the same?

Certain forms of carbon can be transparent to IR. The real dangers lie in what is deposited along with the carbon from the smoke. Candle wax and paraffins are totally transparent to IR and if dirty from partially burnt fuel this can look like proper soot. The potential IR can cause serious damage to your retina. Forget the idea and buy something with a guarantee.

Faulty Filters

The warning signs of a faulty filter are unevenness across the width, bright patches or pinholes. Never use a solar filter that is in any way suspect.

If the worst happens and you can see bright afterimages persisting for more than twenty minutes, you should have your eyes checked over by a professional optician.

And Finally...

What do you do if you are still worried about filter safety? Very simple. Don't use them and stick with a projection method… but this too can have its dangers, as you will see in the next chapter.

Chapter 15
Projecting the Sun's Image

The only really safe way to view the Sun at any time of day with *no* risk to your eyes, is to project its image.

That is not difficult. Indeed, many of us have done this sometime or other in the past using a pair of binoculars, or a small telescope. And that is all you need, the very simplest of binoculars or telescope.

The trick is to line everything up so that sunlight passes straight through on to a screen held behind the eyepiece. The screen can be anything you like – a wall or fence will demonstrate the principle – but a piece of paper or card is best.

Alignment can be quite tricky: you are getting close to the right direction when a bright spot of sunlight flashes across the shadow! Adjust the position of the telescope (or binoculars) until the instrument's shadow is at its smallest on the screen, with the bright spot of projected sunlight in the centre of it. (Do the best you can with the shadow's shape when using binoculars as the shadow is bound to be a weird shape!)

Obviously, this is going to be a lot easier if the telescope is supported on a tripod, but many of us have succeeded on our own. It's a lot easier with a willing helper to hold the screen.

Adjust the focus to get the sharpest image on the screen. (You can "pre-focus" your telescope or binoculars and get it approximately right by focusing on some distant object at the horizon – taking particular care to sight and look in the opposite direction to the Sun!) Make a note of how to turn or adjust the eyepiece to make it move closer to the main lens, and refocus on the distant object again. Now we can realign the telescope with the Sun and screen. The solar image will be almost in focus.

Next move the eyepiece in the direction you found to move it closer to the main lens. Only a little adjustment will be needed to make the Sun's circle snap into sharp focus.

If you are lucky you should be able to see some sunspots, even with a very simple telescope or cheap binoculars. Sunspots (always remember there may not be any of them on any given day!) are a better test of focus. Persevere until they are as sharp as possible, bearing in mind that the edges of the spots will be ragged and the inner portions full of detail and differing shades of darkness. Under some atmospheric conditions (often most of the time in many countries) the atmosphere will be 'boiling' in the heat of the day and it will be extremely difficult to focus on the sunspots with any confidence except in brief periods of "good seeing". Under these conditions, focusing on the solar limb (the Sun's edge) is the only option.

If the image is too small, move the screen further away and refocus. There's a practical limit, which is usually when the Sun's image gets too big to sit comfortably within the telescope's shadow.

How can you get more magnification?

The answer is to make a solar shield. By far the simplest shield can be made with any old rough piece of cardboard. All you need to do is chop a hole in the centre of the board just large enough for the telescope tube to poke through. Take enough care to make the hole nice and snug, otherwise sunlight will peek through the gaps to make "pinholes" – distracting because pinholes are actually an alternative projection device, which will be discussed later.

Another important design criterion is the screen's stability in wind. Make sure the thing is really snug and won't blow away in the gusty windy conditions normal on most days.

A further note of warning is appropriate here. If you use binoculars for projection, make doubly sure (yes, it's a pun, but it may help you to remember…) that the other half or side not being used is blanked off so that the full blast of sunlight does not pass through it and burn or even blind someone. Long experience has taught that the owner of hands holding the screen on which the main image falls will not react happily to a burning image from the other side of the binoculars falling on to his or her skin.

Viewfinders

Most telescopes, even the cheapest, are generally fitted with some kind of viewfinder, usually a smaller "piggyback" mounted telescope, often fitted with cross-hairs. *All such finders and any other optical gadgets on the side must be removed or fully blanked off.* Cap the viewfinder *at the front* with something that won't fall off. There is a very real danger of forgetting that you are observing the Sun and absent-mindedly looking through the viewfinder; if you do this, you could be blinded. Less important in the long term, but very annoying, is the fact that the lens of a viewfinder telescope focuses the Sun's rays directly on its cross-hair sight, which may disappear in a puff of (expensive) smoke. (Yes, you're right: I did say exactly the same thing in the previous chapter!)

Solar projection is one way of getting some practical use out of the cheapest telescopes – the sort that are sold in shops and by mail order as "astronomical telescopes" but which are pretty much useless for the purpose – because the huge light loss becomes a practical benefit in keeping the intensity within bounds. Such "toys" are at least capable of impressing people by projecting a solar image and watching for transiting aircraft if you are unfortunate enough to live near an airport.

A small telescope of around 2–3 inches diameter (50–80 mm) is more than adequate for solar image projection.

Many different "spotting" telescopes are sold for bird-watchers ("twitchers" in the UK) and other viewers of wildlife. These are good telescopes for solar projection, and – having an upright image – can be well used for these more terrestrial jobs when not engaged in the astronomical transit work. Most have a tripod mounting bush, so fixing them to a photographic tripod is simplicity itself. A substantial pan and tilt head (or heavy-duty ball and socket) on the tripod is essential.

Many "spotting" telescopes have an angled eyepiece, usually 45 or 60 degrees to the main body or tube, which makes it a lot easier to see objects on the level without getting into an uncomfortable squatting position. It's a feature that is useful for solar projection because we can work the logic in reverse: if the Sun is high up near the zenith, the practical limit on the size of the projected image size is set by how off the ground you can get the eyepiece. Apart from using a tripod the size of a giraffe (or, come to think of it, digging a hole) you are stuck with that limit.

However, with an angled eyepiece the image can be projected sideways, which solves the problem.

There is another practical advantage. It is possible to project the image into the shade and forget about shielding the telescope itself.

The only downside seems to be in those telescopes with a fixed "stalk" holding the eyepiece upwards, intended to be looked-down-into for terrestrial objects. It takes effort to find the right orientation with a pan and tilt head. Hint: spend a bit of time to test out the configuration first with a hand held trial.

Some astronomical telescopes have a right-angled eyepiece attachment which can be fully rotated in its sleeve. Such a device overcomes the tripod alignment problem and makes them the gadget of choice if one is or can be fitted. Projection at will in almost any direction is then possible. Ensure proper care is taken not to shine the dangerous Sun's image in to someone's eyes.

If safety is your primary concern, then I suppose the safest option is projection directly onto flat ground. That avoids the risk to bystanders, and has the incidental advantages of using a flat surface and keeping the setup compact and all within your control. The debit side is a limit on image size.

For public displays all types of angled eyepieces far outweigh any disadvantages.

Projection on to a fixed screen is then a practical proposition, and the Sun's image can be kept relatively still with skilled tracking or by using a motor drive.

Projection Screens for Visual Observing

Counter-intuitively, it is a great mistake to use a white screen to see the projected image in full daylight. In practice, the contrast is terrible. Strange as it may seem, a blackboard, anything painted matt black or a "grey card" used in photographic calibration will work much better.

The glaringly bright image will show up much better on those screens, and can be viewed for as long as you like in complete safety, without squinting or eyestrain. The best alternative white screen is said to be plain (i.e., no image!) photographic paper. This conventional type of screen works best with the larger, and therefore dimmer, images and when seen in a darkened or shaded area, or a darkened room. Try all options and choose the screen which suits your purposes best.

The great publicity advantage of this projection idea is that any number of people can see what's going on at the same time. It's perfect for public demonstrations and to attract passers-by who are always interested in something like this.

Projection not only guarantees a large number of people can view at a time, but they can also take as many pictures as they like in complete safety. Little or no exposure judgement is necessary as the scene is more or less a normal daylit one. Autoexposure cameras and videos will perform beautifully.

And the general public won't care about the way the image appears. After all, a circle is a circle, or a crescent, during an eclipse.

You should realise that the projected image will be laterally reversed when it comes straight out of the eyepiece on to a screen. That can play havoc with scientific studies if not taken into account when plotting sunspots or predicting where on the limb a transit will start (or end). It is also worth bearing in mind that it might be possible to read lettering on aircraft at the higher magnifications and these will look most odd when shown as a final print! To get the print looking right it might be necessary to have the original "flipped", by printing the picture with the negative "upside-down" in the enlarger, or by using appropriate software like *Paint Shop Pro* on a scanned photo or negative.

Sunspot monitoring conventions allow for this oddity and the true solar orientation makes it clear which side is which. Spend a bit of time to clarify the orientation in your own mind when projecting the image sideways, particularly at times of day well away from local noon. The first clue is given by the direction in which the image moves across the screen with an undriven mount. Make a note of this on the projection screen with an arrow.

The biggest mistake a beginner can make in solar projection is to jump straight in with a modern mirror-optics astronomical telescope. Even a small mirror optic sold for "spotting" purposes is not a good idea either. There are two sound reasons for this, one scientific, the other practical.

The first word of caution with mirror optics and projection involves any use for scientific purposes because of this image orientation problem. The image may be rotated and reversed in a number of awkward and very interesting ways according to the optical system. It all depends on the number of reflecting surfaces, and how they are arranged. Some reflecting telescopes confuse the issue further with added lenses before the eyepiece.

Then, when you add into the equation right-angled viewers the whole thing becomes quite confusing. Some right-angled viewers laterally invert the image anyway, others are designed to yield a normal or terrestrial view and so on ad infinitum. When the Sun is in its most awkward orientation mid morning or afternoon, it's best to give up trying to sort it out and just accept what is on offer.

Before using the technique to record sunspots or anything remotely scientific you must make proper notes to check the orientation, which is unlikely to be as seen in real life, that is, as it looks with a naked-eye filter used at the same time. The most helpful advice is a bit tedious but it will pay off handsomely in the long run…

Wait until night-time and find a nice bright neon-lit hoarding or other sign with writing on it and project the image. A low-power eyepiece is much better than the higher magnifications used on the Sun. The writing will tell you how the projected image is oriented. It might be inverted or not, a mirror image, and so on.

Alternatively, you could use the Moon if you are confident enough to sort out the crater topography.

Now, on to the second and more serious reason why a mirror telescope is not a good idea for solar projection.

The sheer size and consequent light-grasp of today's SCT's make them dangerous. Read, learn and inwardly digest the potential hazards discussed in Chapters 13 and 14 concerning hazards in general. Even with a small catadioptic telescope and a "safe" eyepiece there is a large amount of heat going into the tube, and this will have a devastating effect on focus stability, something mirror optics are very sensitive to.

All in all, there is an "open and closed case" for sticking with refractors, with their inherently better focus and temperature stability.

Specialist Solar Telescopes

The professional astronomer's answer to solar projection is to create a permanent structure with mirrors. This time we are not talking about curved surfaces as in a telescope proper but optically flat surfaces to reflect the light into a small static telescope which projects the image onto a screen. Depending on the number of mirrors used, the system is called a heliostat or a siderostat. There are quite a number around the world and some in amateur hands.

The biggest of all is the McMath instrument on Kitt Peak, Arizona, USA. This has a large tunnel underground where the air temperature is stable. Another large design uses a tower with the mirrors at the top, directing sunlight straight down into the observation room underground. The best known of these is the one on Mount Wilson.

Amateurs can try an alternative and greatly simplified version of this reflecting mirror idea.

No telescope in involved! This is becoming widely recommended as a method for solar observing during times of eclipse. All you need is a little flat mirror (of the sort used in cosmetic compacts). Only a small area is needed – less than an inch (25 mm) – across. Mask it down to this size. The solar image is simply reflected on to a shaded screen several yards (metres) away.

If you mount the mirror on a CD case it gives a degree of altazimuth control.

The idea works very well to demonstrate the overall shape of the Sun during times of eclipse, but of course the resolution is not really up to the job we need to spot very small objects transiting the disk. The image scale or magnification is also somewhat poor, giving a solar disk around 4 inches (10 cm) across at about 10 yards (10 metres). The edge of the Sun is visibly fuzzy. Nevertheless, the idea is a good way to start, and will work to spot aircraft movements.

However, there is a natural way of projecting an image and you don't need any apparatus to do it. Leaves on trees or holes in any obstruction will allow sunlight through to form a "pinhole" image. The dappling sunlight under a tree breaks up into delightful crescents during an eclipse, and become much more pronounced just before and after totality. On a normal day we pay no attention to the solar image because it is circular and undistinguished. You can make the effect much more useful with a single pinhole. Make one by punching a small hole in a piece of cardboard, and hold it at arm's length from a screen.

As with all projection methods the contrast is too poor on a white screen in daylight, and much better effects are seen when the screen is grey or black. The longer your arm, the bigger the images!

Some of the best pinhole images can be seen in a darkened room with a small hole in an opaque or semi-opaque window blind. The image will appear on the opposite wall or floor. That is precisely the way the Sun was observed in Mediæval times, and how sunspots were studied before the telescope came into wide-spread use.

Resolution, image-definition and contrast are surprisingly good by pinhole projection into a darkened room. It's a useful answer for long-term studies of the Sun if the room's orientation is correct.

Chapter 16
Observing Transits

The thrill of seeing a transit for the first time is what really matters, not the technique used....

Many observers can switch between naked eye observing (and maybe drawing their results) and photography, and are quite happy doing so. It is less common for a committed naked eye observer to get fully immersed in photography, but the converse is almost inevitable since visual observing is necessary for focusing.

For convenience the differing techniques and equipment needed for visual observing or recording are split into two chapters. The skills for visual observing or photography are so inextricably intertwined that each reader will rapidly come to the right balance for their own financial or time constraints.

Telescopes

So far, we've talked more about what kinds of telescope *not* to use than what are the best types.

The supreme art of transit observing is detecting relatively small objects at the very edge of a larger disk. When the transit is of a sunlit satellite across a major planet, light gathering power is of lesser importance than resolution and contrast. The two factors are not mutually exclusive, of course, and as in a lot of everyday life, the ultimate resolving power of a large telescope is going to be better than that of a small one, all else being equal.

If this seems to go counter to what I said before, remember that a telescope can always be stopped down with a mask over the front aperture, and a large-aperture telescope is usable for normal "night-time" astronomy as well.

The next consideration is what type of telescope to buy – for solar work – and is a little easier to decide.

- For visual-only work   The refractor has to be a prime choice.

- For conventional   With a silver halide medium, photograph   a fast focal ratio mirror optic is the only practical option on a tight budget.

- Electronic detectors,   Opt for a refractor to get the all types   best quality primary images.

The superiority of refracting telescopes for this purpose is very clear from this brief list, and if expense is no restriction the modern fast breed also score hands down for conventional photography. Why should this be?

Light-gathering power is the single most important factor in most astronomy where deep-sky objects are being observed. Normally this means a huge reflector, because reflectors (or catadioptric telescopes) are the only ones capable of being made large enough.

But light-gathering power is very much a secondary consideration in transit work because there is often too much light from the primary body. Remember that Jupiter's disk is a sunlit object! Jupiter is so bright that it can often be seen with the naked eye through moderate cloud when no stars are visible (as happened the evening this chapter was being written). Jupiter's brightness is an appreciable fraction of the daytime landscape outside – comparable to this page if you are reading under artificial light!

With so much light about, *contrast* is much more important than anything else, and that's where refractors score heavily, aperture-for-aperture. The central obstruction lowers contrast, and "ghost" images of the secondary mirrors or baffles degrade contrast still further when viewing brightly lit areas. This second effect is precisely the last thing we want to encounter when looking out for transient shadows and the like!

Some of the reasons for this choice should become clearer when we review the characteristics of each type in more detail.

Mirror Optics

All modern astronomical mirror optics from reputable manufacturers are excellent, and with a minimum of care will yield years of useful life before needing attention.

The basic problem with all mirror optics – even with sealed tubes – is the fall-off in contrast in the course of time. It's inevitable (and much more rapid) with an open tube of the home-made variety.

Deterioration of the metal coating on the mirrors has two consequences. The most important for us is a dramatic decrease in contrast. Reflecting telescopes already have a lower contrast than the same aperture of lens optics, mostly because of the central obstruction in the primary mirror.

The second consequence is of less importance for transit work, but is worth noting. All common mirror optics have at least two reflecting surfaces, and as the mirror surfaces degrade (from surface encrustation, tarnishing, or too much cleaning) the reflectivity will drop.

For transit work we usually have a glaringly bright object to look at, but you should still be aware of the factors that affect performance. Check for the effect when there is a clear day and see if a dark patch can be picked up against the blue sky as the eye moves across the eyepiece. If there is a dark patch moving about as your eye scans, beware of spurious observations when there is a lot of light entering the telescope tube.

Lens Optics

At one time it was almost impossible to get a refractor with anything approaching the mirror's ability to focus all parts of the spectrum properly. Chromatic aberration (the inability to focus light of more than one colour in the same plane) is reduced by using combinations of lenses made from different types of glass, of which the *achromatic doublet* is still the most popular.

Specialised glasses and fluorite have made dramatic improvements in recent years, and some very respectable reductions in chromatic errors started to appear in commercial items by the 1970s. The impetus needed for faster and faster zoom lenses for cameras hastened the next step which was aspheric lens surfaces. The improvement in chromatic corrections has been almost as remarkable as the parallel one of lens "speed" or focal ratio. An aspheric fluorite telescope better than f/5 and a respectable 100 mm+ aperture is (almost) affordable to many, and may be a better investment than mirror optics twice that diameter.

The final improvement in lenses was the improvement in anti-reflection coatings. They increase light transmission and – again, important for transit work – reduce reflections and thereby improve contrast. Transmissions approaching 100% are commonplace.

Conclusion

Our (the authors') conclusion is that refractors, of modest aperture, are likely to be the best bet for transit work. They tend to have better contrast and, with modern lens design, chromatic aberration is not significant. For specialist observing of transits, a large aperture is not usually needed.

However, your choice may be conditioned by the fact that you want a general-purpose instrument, and nothing here is meant to imply that SCTs are in any way *un*suitable.

Alternative Optics

One of the interesting spin-offs from the rapid progress in electronic imaging has been the large number of "obsolete" camera lenses coming onto the second-hand market at bargain prices. Many of these spare lenses are of excellent quality and can be pressed into service as telescopes. There are a number of practical advantages in this idea.

The price is the most obvious one. A simple refractor can be made at a fraction of the cost of a comparable astronomical one of the same aperture. Furthermore, the optics will be well corrected for optical aberrations, and of a much faster f/ratio than most telescopes of the same focal length. Nearly all of the lens manufacturers make special eyepiece adapters for just this purpose of turning their lenses into very serviceable spotting telescopes. Independent manufacturers specialise in making custom adapters to turn lenses into proper telescopes with astronomical-style eyepieces.

Some obsolete lenses are actually modified telescopes in the first place, and the classic examples of these are the *mirror lenses*.

It is often a matter of moments to convert them. Perhaps the best known are the Pentax 42 mm screw lens fittings still turning up from the former Soviet Union. All the Pentax screw gadgets are widely available, since some cameras are still being made with that fitting. The most versatile range of lenses with superb adapter optics is the Tamron series with the "Adaptall" fitting. These adapters are just right for terrestrial purposes as they do not give laterally reversed images, something which has to be taken into account for astronomy.

The prize lenses have to be the "Pentacon" variety made by Zeiss in the former East Germany for their 6×6 cm format cameras. These are ultra-rugged, and because they lack all the modern autofocus and other gimmicks, they appear on the second-hand market at really low prices. The best of the lot is the 500 mm f/5.6 which often comes with a Pentax 42 mm rather than its usual P6 adaptor, making it an easy job to fit eyepieces or almost any recording device you like.

For a fraction of the cost of a conventional 4-inch, this brute is a full $4\frac{1}{2}$ inches (118 mm) diameter, and can be stopped to down to f/22 in stepless increments, much simpler to use than neutral density filters. The optical quality of an optical system designed to cover a full 6×6 cm, when used only with an eyepiece or a 35 mm camera at the highest definition at the centre, makes this an ideal beginner's entry on a budget to any form of astronomy (see also Chapter 11 page 83ff).

Amateur Solar Telescopes

As we have seen in other chapters, transits of Venus and Mercury across the Sun are exceedingly rare but we do have some coming up in the near future. There are already a large number of solar observatories in action and these could be rescheduled without any fuss if they happen to be in the right place at the time. Many are not.

Portable telescopes used for eclipse studies also do not need any modification provided the image scale is large enough. Now is the time to consider all the options and get ready. Constructing a special portable transit telescope is not always possible and some of the authors' experience might help here.

The prime need is flexibility with good resolving power under extreme heating conditions – unlike during a solar eclipse, we are dealing with full sunlight, all the time. Mirror optics do not stand up too well unless special care is taken to shade as much of the telescope as possible. We simply do not want the huge light gathering power of a large mirror under full sunlight, which throws us back on first principles of quality optics – which in our view equates to good quality lenses.

Right-angled viewing is also needed, with the option of photography whenever possible. A partially silvered prism serves both purposes, and the final construction is the "Poyser" design (named after Ian Poyser who made the prototype). This has also proved its value at solar eclipses and sunspot monitoring, its main duty. It is about as flexible a piece of equipment as any, and allows direct viewing, photography or projection through either eyepiece and any combination as circumstances change.

Figure 16.1. The huge scale of the McMath solar telescope is obvious from this scenic view.

Figure 16.2. The heliostat of the McMath solar telescope is capable of projecting a solar image down into this tunnel where observing conditions underground are much more stable.

Eyepieces

Eyepiece technology has also come on by leaps and bounds in parallel with telescope design, to the point where a cheap new one performs infinitely better than an expensive old one of a few decades ago. Anything sold by a reputable dealer will come with a guarantee and be fine with any type of telescope. The choice then becomes as much personal as practical.

The long observing periods needed for unpredicted or unpredictable transits demand comfort as a prime consideration. High magnifications might be necessary but the lower powers could be more useful in the long run in reducing eyestrain. Good advice is to use the most comfortable for routine monitoring sessions, and then switch to any higher powers at times of high excitement. *Parfocal* eyepieces are useful here – it just means that you can switch from eyepiece to eyepiece without having to re-focus.

The simpler and more robust the lens mount is, the better!

Personal preference has to be the decider. Always try out a new style of eyepiece "on approval" if you can, or borrow one from a friend before purchase. Eyestrain has been identified as a prime culprit when observers throw in the towel too soon. All too often it is poor choice of eyepiece rather than a poor telescope at root cause.

Plossls, and similar lenses such as Naglers, with what is known as good eye relief (your eye doesn't have to be right up against the lens), are very popular because of their splendid views and reduced eyestrain. Anything with a large exit pupil has to be good news.

Observing Tips

Because there have been reports of bright spots seen on planets' disks during solar transits, particularly of Mercury, heed the warnings above when observing. Mirror optics can give ghost images, the most likely cause of these spots. However, many of the older refractors without good colour correction can also give some horrible contrast effects, or false colour focus spots in the centre.

Because of our experience at the Sigma Sagittari occultation in 1981, optical quality does seem to be the most desirable feature in the observing equipment.

Shadow Transits in Practice

Shadow transits on the major planets do not give a solid "blob" of blackness with a clean edge as you might expect. There is always a fuzzy border which is entirely dependent on the distances between the two transiting bodies and the illuminating source (the Sun), and the sizes of the three objects. The geometry is such that there will be a central dark portion – the umbra – surrounded by a lighter region – the penumbra. We see much the same thing at solar and lunar eclipses.

Table 16.1 now shows why we have emphasised the importance of telescope quality, and specifically the contrast.

Without a good telescope, the shadows will just appear as an indistinct blob with no shading from umbra to penumbra. Each satellite does have its own

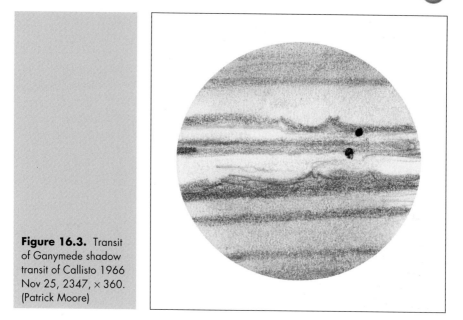

Figure 16.3. Transit of Ganymede shadow transit of Callisto 1966 Nov 25, 2347, × 360. (Patrick Moore)

Table 16.1. Major planets and recognising their satellite shadows

Satellite	Satellite disk	Umbra	Penumbra	Ratio u/p %
		average arcseconds diameter from Earth		
		Jupiter at 5 AU		
Io	0.97	0.81	1.14	70
Europa	0.84	0.55	1.12	49
Ganymede	1.14	0.92	1.88	49
Callisto	1.28	0.41	2.14	19
		Saturn at 9 AU		
Titan	0.79	0.62	0.97	64

distinctive appearance, which can almost be likened to a signature. The signature is a quick and easy way of identifying or confirming which satellite is involved.

The table shows that the shadow sizes relative to the planet's disk and areas of umbra and penumbra are very different. With some experience the "hardness" of the blob becomes very useful in confirming the observation. Io's shadow is invariably much "blacker" than Callisto's greyish hue. Now you can see why.

It's a useful piece of observing know-how under bad seeing conditions. It can even be used to test telescopes suspected of less-than-optimal quality.

All the other satellites cast shadows smaller than the 0.2 arcsecond resolution limit in an average atmosphere/seeing. Most telescopes are capable of resolving better than this seeing limit. That is why a shadow transit of these minor satellites can be glimpsed in periods of good seeing.

The four Galilean satellites can also be identified from their appearance when in transit themselves, with or without their shadows.

Io invariably appears grey when it can be seen at all. Most observers often regard this as an "invisible" transit in the smaller equipment.

Europa is almost as difficult and is normally classed as "invisible".

Ganymede is invariably very dark.

Callisto is also dark because like Ganymede it is a lot larger than the other two but also of a lower albedo. That makes it a noticeably fainter magnitude than the other three when seen against the night sky.

Eyestrain

Eyestrain problems are much more common than most astronomers will admit.

It hits at the core of observational astronomy, but to admit it risks comments that you are no good at it or at the very least fast approaching senility! Severe eyestrain can usually be traced back to bad tuition, or a complete lack of it.

The best training you can get is to start with the opposite end of optical instrumentation, with microscopes.

With most telescopes, the focus is computed so that the image falls at the eye's most restful point. In our primæval past, human eyesight focus and rest point evolved to be most efficient at a distance around 30–90 feet (10–30 metres) away. There was a fundamental survival value in this distance – it's the distance beyond which we had a chance of getting away from predators intent on eating us! At night-time, depth of focus is minimal with the eye's pupil wide open.

That's why telescope focus is set around this distance (or infinity) with the best of intentions for comfort. Fortunately (or unfortunately, depending on how you think about these things) humans have not been eaten in any quantity for millennia, and correcting lenses for short- or long-sight compensate well enough.

The result – many people find their corrected focal point is not comfortable during a long session, either with or without glasses!

How often do you see an observer squinting through an eyepiece and quite likely making the situation worse by closing the other eye? This is absolutely the wrong technique.

Microscopists know about this problem and train people to keep both eyes open. The reason for this is simple. Microscopes are computed to give the maximum image size, and that is done by focusing at the closest point possible – the normal reading distance of around a foot (25–30 cm). We all know that a lot of reading strains the eyes and it's the same cause as microscopy. Close focus is an unnatural distance for restful vision.

People who screw up their eyes, squint or close the other eye soon learn all about eyestrain, but by keeping the other eye *open*, both eyes can settle down better. It's surprising how quickly the brain accommodates to looking at the image in one eye only "on demand".

Long sessions at the telescope eyepiece are that much more restful and hence enjoyable, once the trick is learnt. These long sessions waiting for a transit can be filled in with a "weather eye" open for transient events elsewhere in the sky. Sometimes a schizoid personality is most useful!

Some observers do find it too difficult a trick to learn and use an eyepatch, or a blanking bar fixed to the eyepiece mount. No matter, the important thing is to keep both eyes rested as much as possible.

A final eyestrain tip. Whichever technique you use to help relax your eyes, light pollution is something we can do without, but seldom avoid completely. A shroud over the head looks ludicrous but can be amazingly effective in these circumstances. What we need is something like the old-fashioned black cloth used in Victorian times to focus camera screens properly!

The improvement in the ease of seeing the fainter objects is quite dramatic as dark-adaption improves. It all adds to contrast enhancement. And the enjoyment of spotting a transit for the first time (see also Chapter 11, page 83ff).

"Murk"

Much has been made of the contrast factor when reviewing telescope design. Stopping down is by far the simplest way of improving any telescope in that respect, certainly if it has mirror optics. There is more than enough light around with the major planets for that dodge to work well.

Unfortunately all the major planets have been sitting in the summer constellations for many years now and anyone in the northern hemisphere much above about 40°N had to wait patiently for some time each night before the disks rose above the horizon murk. The situation is improving, but even Jupiter is not fully out of the murk until a frustratingly long time after rising. Views for astronomers in many locations are hampered to brief moments of clear seeing, much more than will be the case in the early years of the next century. The contrast through the murk can leave a lot to be desired, particularly when chasing shadow transits.

Stopping down comes to the rescue again.

Mirror optics often improve considerably when stopped down and a useful rule-of-thumb is to mask or stop down to about 25% (equating to two stops in photographic terms). Note, this is 25% of the *area*, not of the diameter. Sometimes a little more stopping down might be even better. The contrast improvement can be little short of dramatic and improve quality observing time beyond all recognition.

Refractors also benefit from this simple technique – arguably even more so. In this case the benefit is in reducing the remaining chromatic aberration. It seems in practice to make the true colour of objects more easily visible.

● Conclusion   In conditions of poor seeing conditions always consider stopping down a telescope, whatever the type.

Chapter 17
Photographing Transits

Telescopes

The previous chapter listed some very important design features:

- For visual-only work   The refractor has to be a prime choice.

- For conventional photography   With a silver halide medium, a fast focal ratio mirror optic is the only practical option on a tight budget.

- Electronic detectors, all types   Opt for a refractor to get the best quality primary images.

The resolving capability of any telescope is described in any textbook, or it can worked out from first principles with very simple equations. That is rarely needed in practice as the essential data have been published in many places.

What we find most useful is the intriguing coincidence in the case of Jupiter.

At its mean distance from Earth, the apparent diameter in the eyepiece is then 40 seconds across. A focal length of 5m at prime focus gives an image so close to 1 mm diameter at that mean distance as to be acceptable as exactly that size. At opposition, Jupiter is just marginally larger than this "rule-of-thumb" measure.

A 1 mm image is extremely small and yet a 5000 mm (over 15 feet) focal length telescope is much longer than even most of the commercial SCTs on sale. It's easy to see why prime focus photography is so disappointing!

Film resolution is just about able to show some detail, but CCDs are simply not up to it at an affordable price (yet). That is why direct views with the eye through a magnifying eyepiece still have a place as the magnification can be pushed to the absolute limits the seeing will take. The superb computing and processing software in our brains take care of the rest and can respond fast enough to analyse the fleeting glimpses presented.

It also explains why CCD technology comes into its own and scores when images are enlarged through some projection technique. The image brightness intensity falls off as the square of the magnification, and every light quanta can be registered fully. Silver halide films tend to bottom-out long before maximum image scale is reached.

Table 17.1 lists the prime-focus image sizes of the main objects of interest in transit photography. It illustrates in passing why the older-fashioned long focus refractors still have a place. A 1000 mm focal length is at the top end of camera telephoto lenses, and typical

Table 17.1. Diameters of astronomical subjects at prime focus

Subject		Visual seconds	fl 1000mm mm	5000mm mm
Mercury				
	May	12.9	0.06	0.32
	November	9.8	0.05	0.25
Venus		65.2	0.33	1.63
Jupiter (mean)		40	0.2	1.0
	opposition 1999	49.8	0.25	1.25
	2000	48.5	0.25	1.21
	2002	47.1	0.24	1.18
	2003	45.5	0.23	1.14
	2004	44.5	0.22	1.11
	2005	44.2	0.22	1.11
Saturn (mean)		17.9	0.09	0.45
	opposition 1999	20.3	0.10	0.51
	2000	20.5	0.10	0.51
	2001	20.6	0.10	0.52
	2002	20.7	0.10	0.52
	2003	20.7	0.10	0.52
	2005	20.6	0.10	0.52
The Sun (mean)		1920	9.6	48.0

of many of the popular spotting telescopes used in amateur circles and for bird-watching.

The table confirms that both Mercury and Saturn are severe tests of resolving ability and place a premium on contrast.

Cameras

There is still a place for conventional photography in transit studies, because the exposure times cannot be as extreme as deep-sky and similar esoteric work.

Having said that, I wonder how long cameras will survive in view of the manufacturers building in more and more "features" to compete commercially... it results in ever more complexity and just adds to the expense. "Auto-everything" becomes quite counter productive in astrophotography, especially when a camera relies on tiny batteries that are being drained all the time the shutter is open!

The ideal camera for all forms of astrophotography has to be a basic design with an ability to perform the following simple functions:

Brief time (marked "bulb" on old cameras), or
Time exposures
Manual shutter or low battery consumption
Cable release
Mirror lock

Anything else, (apart perhaps from a motor drive) can be more of a liability than an asset in the dark. And the more expensive the camera, the less inclined you feel to risk it in a damp observatory or damper field at night.

A *mirror lock* facility is almost a prime requirement with the extremely long focal lengths needed.

Zenits and similar inexpensive camera types, which come into their own for long deep- sky exposures, tend to rule themselves out for transits because the alarming shuddering as the reflex mirror slams up can take several seconds to die down.

Most major camera manufacturers made a model or two with mirror lock in their manual-mode range. Consult a good second-hand dealer for suitability as they reappear on the market, if a new Nikon seems to be beyond your price-range just now.

The only sensible alternative with the Zenits and their ilk is the "Mexican Hat Trick" (have your Mexican helper hold his sombrero to cover but not touch the front of the telescope, open the shutter, wait for the instrument to stop vibrating, time the exposure by removing and replacing the hat, close the shutter). Judge the exposure time as best you can.

Practice with the equipment is absolutely essential until exposure times can be guaranteed under all the range of seeing conditions experienced in the normal weather at the observatory. There are specialised astrophotography books to assist.

Many modern cameras lack a cable release; others cannot be set in manual mode for exposure time. Nearly all burn up battery power and are useless without it. Always have spare camera batteries with you.

Films

Silver halide films have undergone a major transformation. Before the mid-1990s the pundits were predicting the death of "ordinary" film by the late '90s. Almost the opposite has happened, and film sales have held up remarkably well because the film manufacturers put in the necessary effort to improve the product to compete with electronic imaging. At the time of writing this, all but the most expensive digital cameras have visibly inferior resolution to film, even for modest enlargements.

The main improvements in film – latitude (the ability to record detail in over- and under-exposed parts of the negative), improved "grain", and vanishing reciprocity failure, have been a boon to astrophotographers. There are now many film types designed for "snap-shotters" which outperform the specialist astronomy film of the early 1990s! And of course, these new films are available in colour...

All the major manufacturers make colour films guaranteed in the 400 to 1000 ISO range with a fineness of grain (granularity) that was undreamt of a decade ago. They all also make monochrome film – with the same exposure latitude – which is compatible with the same colour processing chemistry. Ilford have now brought out a new monochrome Delta film rated at 3200 ISO to compete with Kodak's T-Max 3200. Reversal colour films are also following the same interesting trends.

The range is expanding, not contracting, and a suitable film can now be found for almost all exposure conditions. Consult the literature and astronomy magazines for the latest reviews before giving up on "wet" photography for digital imaging..

Another factor in the revival of interest in silver halide films is the very same as that which makes CCD imagers so attractive. All the same electronic scanning and processing technology can be applied to conventional film; your processing house can scan your negatives at high resolution directly on to a CD-ROM! This is quite cheap to do either when the film is processed, or afterwards. You can then computer-enhance your images, see Figure 18.1 (p. 157).

Films also constitute an archive in their own right, not an electronic artefact!

Digital Cameras

Interesting to try, but by no means true astronomical CCD cameras, digital camera technology is advancing rapidly. So far no clear-cut industry standard has emerged. All the major camera makers are involved, along with a number of new players. Each month some new development catches the fashion, and that is likely to continue until definitive standards are established.

If you want to have a go, it is a case of experimenting, but don't expect too much. And as with CCD imagers, buying a camera is not the end of the matter. A photo-quality colour printer is needed, and usually a PC to get the best out of the images. Software to process the images might be included in the computer package, and many offer enough of the basics to start, but be prepared to spend more on better software with the advanced features you will need for astronomy.

Apart from their resolution (or rather lack of it) compared with that of film or astronomical CCD imagers, the limiting feature with many digital cameras is a complete lack of adaptability to high-quality optics or a telescope. Remember, digital cameras are simply not yet good enough to satisfy the quality demands of serious photographers (never mind about astronomers), and rely for their popularity on instant access and hours of endless amusement "improving" the images via the computer and software. For scientific use, these cameras are not yet an option.

As the resolution improves, as it undoubtedly will, it is likely that situation will change or even reverse. Digital cameras will one day take over from CCDs for a lot of routine astronomy where the subject light intensity is high. Transits of the major planets will be a logical case in point.

Video Cameras

If you are an astronomer interested in transits and own a camcorder, you might be in for a pleasant surprise.

Generally speaking, domestic video recorders are not sensitive enough for astronomy. They will pick up some of the brightest stars but not nebulosity, because they do not (usually) have a cumulative storage capability. However, when dealing with the Sun, or Jupiter and Saturn with a fairly large telescope, the light intensity can be sufficient for some of the newer and more sensitive models to capture a surprising amount of detail.

Experiment to see what you can do. It is certainly possible to record spectacular videos of the Moon. Finding the right sort of adapting mount could be a challenge, because domestic video cameras do not often allow lens changes or removal.

Commercial cameras for broadcast quality (or near-broadcast quality) are a lot more versatile and usually do allow direct fitting to alternative lenses like telescopes, but usually they are *heavy*.

The main practical advantage of this option is the ability to play back the final result through any domestic monitor or TV – great for public shows!

Image Intensifiers

Image intensification was the only alternative to visual and photographic recording open to astronomers for a very long time. When CCD technology first appeared on the scene, image intensifiers largely dropped from sight except for variable star work. The technology seems to be enjoying a revival and could replace CCD imagers in some applications. Transits are a case in point.

Image intensifiers are available in High Street camera stores. They have become fashionable as "night scopes" for surveillance and similar activities. They work quite well, but aren't the best you can get. The most effective (military specification) units are very expensive and may not be easily available. However, some of the better ones do "leak" on to the domestic market and have been applied to specialist astronomical observations. Meteor observers have found them particularly useful and used them a lot more in recent years.

Using this kind of image intensifier is simple. A lens (telescope in our case) is fitted to the image intensifier which is then turned on. A screen, viewed through the intensifier's eyepiece, displays a vastly intensified image of the scene.

For astronomy an ordinary video or similar cheap camera or recording device can be attached to the viewing screen. The video is looking at the bright image and so can record something quite beyond its non-enhanced capability.

This hybrid technology is set to become much more familiar. Commercial items are now being sold with the latest *Generation-3* image intensifiers specifically made to fit into a normal telescope eyepiece holder. These so simplify the equipment that the price becomes highly competitive with CCD alternatives.

The obvious advantages of this combined technology are extreme sensitivity and a record in real time which can be played back anywhere. All you need is a modern domestic TV set. After that, all the fancy stuff can start, with frame grabbing (usually an extra card in your PC) and then computer manipulation and enhancement that will keep you fascinated for hours...

CCD Imagers

We have left the most popular technique today to last because it is not always the cure-all that it is billed as.

It does give the most sensitive results (intensifiers apart), and the equipment is designed to fit onto any optical design you like. For those two features alone CCD technology has to be the way to go for many observers.

Our reservations centre around the expense and dedication needed: viewed in a proper context, CCD cameras are not expensive, *but* CCD cameras have tiny chips, which in turn have a very small field of view.

In the context of transit observing, there is seldom much of a problem finding the major planets. But the image-scale is such that a CCD camera almost demands a computer and autoguider, and the expense of the combination has to be included.

Unless you are already using a CCD imager, our advice is only to take the CCD option *once you have some experience with simpler equipment.*

If deep-sky observations is the main theme, with filling in with the occasional transit or other planetary observation, then the best CCD camera you can afford is the way to go. If you get to use your telescope only occasionally, or want to observe the Moon, planets or Sun, (or transits!) along with the brighter celestial objects, then "wet" photography may well be the best way to make your images.

Visual Observing

Never forget that the cheapest option – direct visual observing – is unrivalled for "star gazing" and revelling in the wonder of the night sky. It's also best for catching those fleeting moments of excellent "seeing" that suddenly and briefly reveal extra detail in the planets, Moon or even Sun.

Chapter 18

Data Capture and Manipulation

The previous chapter contains most of the information needed to identify objects in transit. Table 17.1 identifies relative sizes and characteristics. Now that we have some interesting observations "in the bag", how do we access the data?

Images on Film

Both negatives and positives (slides) remain popular and all scientific studies should be made using these originals. Unless you are really skilled in photographic processing, prints are only suitable for the purpose of illustration or for decorating a wall: the loss in quality and/or resolution with most commercially produced prints is simply abysmal!

A good quality light-box is always a good investment. Spotting faint detail by holding a negative up to the light is never easy. Trying to hold a magnifying glass at the right distance becomes a test of patience rather than anything else, and is not advised. The difference in checking ease with a light box and a good quality loupe is so marked that nothing else can be recommended.

Checking by light-box or projection is the familiar stand-by. The modern alternatives use all the tricks and technology of electronic recording.

- Develop and check films at the first opportunity. Transit observations are often important news.

Computer Enhancement of Images

As described, negatives (or positives) can be transferred to CD-ROM by the processing house; CCD images are, of course, by their nature digital.

Although the hardware technology is still changing rapidly (mostly getting faster at the moment), the basic enhancements are merely an extension of familiar photographic techniques. Computers only need software capable of contrast and colour manipulation to get you started.

A good software package (which need not be too expensive) will have many refinements for image processing. *Paint Shop Pro* is one of the more popular packages at the moment, and includes useful (to an astronomer) functions such as *sharpen, sharpen more, unsharp mask, despeckle,* and *edge-enhance,* plus a comprehensive range of controls for *colour, contrast, hue, gamma correction* – you name it!

The finished product can be enormously "improved" over the original. Computer enhancement is much easier, more flexible, and far less costly than trying to do the same thing in the darkroom.

Modern inkjet printers can produce results that, to the naked eye, are indistinguishable from photographic prints.

The important thing with any type of computer enhancement is to maintain even more meticulous records than conventional methods, and *keep the originals untouched,* whether film or digital. Archivists in the future will still want to do their own research in ways we cannot know. Make back-ups on a separate medium (floppy disk, etc.).

Remember too that an electronic record might not be as permanent as the written methods, and there will not be a hardcopy original in the same way as a film negative. In a few year's time, the equipment to read your original may no longer be operational. What do you think would be the accessibility of data stored, only 20 years ago, on *Commodore 64* computer tapes?

a

b

Figure 18.1. a A "straight" photographic colour print from a reputable processing laboratory. **b** The same negative, scanned onto CD-ROM (by the same processing laboratory), then slightly enhanced using *Paint Shop Pro 5* (contrast correction plus the "sharpen" algorithm) and printed on a inkjet printer. (John Watson)

Video and Other Methods

The great advantage of any video or camcorder is that the result conveys a feeling of being present in real time. Most of the machines have a timing facility built in which can be as precise as the frame speed, usually 25 frames per second. If the image brightness and sensitivity are good enough, higher speeds can be achieved.

Therein lies a problem: and we can learn from the experiences of meteor observers.

More advanced meteor observers regularly study video tapes generated from image intensifiers. That can mean ploughing through several hours of tape, watching for sporadic events. At least with transits the timing should be known reasonably accurately!

Connect the video output and adjust the monitor screen to maximum contrast. Adjust the colour and brightness levels for comfort and watch in a darkened room. The same rules apply as for a primary visual observation.

If it is a feature of the original recording, set the video to a slow shutter speed with a maximum gain.

You can't really concentrate for more than 15 minutes or so, and 30 is an absolute maximum, even for experience viewers. More often than not there will be things missed during the first run and that is why the rule-of-thumb accepted by most devotees of the technique is that it takes at least three times the duration of the original observation tape to identify all interesting things. Commercial scanning programs already exist, and this 3 to 1 monitoring chore should rapidly become a thing of the past.

The main snag with video tapes is the relatively low resolution of the VHSS standard, less than 399 lines in practice. The more modern Hi8, Super VHS and the obsolete Betamax are somewhat better.

The video option sometimes allows the direct addition of radio timing signals derived from an atomic clock (transmitted in most countries at a low radio frequency). Some economies are possible in making a final (hard-copy) output if computer enhancement is not available. Frame freezing does allow a conventional photograph to be taken of the screen. A portrait lens, say a 135 mm, on a 35 mm camera corrects many perspective distortions present in a standard lens. Use a longish exposure to even out the screen flicker. Two to four seconds is recommended with a slow film for best contrast. Be guided by the camera exposure meter.

There are tape analysers on the market already. Within a very short time, digital TV technology will find wider applications in picking up transits and other transient phenomena.

Here are some practical tips:

- Never make an original record in long play mode.
- Always use the best quality tape you can buy.
- Never re-use a tape for an important observation.
- Always activate the non-erase option on the master tape immediately you take it out of the recorder.
- Copy the master and work from that.

Appendix

ELONGATIONS OF MERCURY, 1999–2005

Western
1999 April 16, August 14, December 3
2000 March 28, July 27, November 15
2001 March 11, July 9, October 29
2002 February 21, June 21, October 13
2003 February 4, June 3, September 27
2004 January 17, May 14, September 9, December 29
2005 April 26, August 23, December

Eastern
1999 March 3, June 28, October 24
2000 February 15, June 9, October 6
2001 January 28, May 22, September 18
2002 January 11, May 4, September 1, December 26
2003 April 16, August 14, December 9
2004 March 29, July 27, November 21
2005 March 12, July 9, November 3

ELONGATIONS OF VENUS, 1999–2005

Western
1999 October 30
2001 June 8
2003 January 11
2004 August 17

Eastern
1999 June 11
2001 January 17
2002 August 22
2004 March 29
2005 November 3

INFERIOR CONJUNCTIONS OF MERCURY, 1999–2005
(* = transit)

1999 March 19, July 26, November 15*
2000 March 1, July 6, October 30
2001 February 13, June 16, October 14
2002 January 27, May 27, September 27
2003 January 11, May 7*, September 11, December 27

2004 April 17, August 23, December 10
2005 March 29, August 5, November 24

INTERIOR CONJUNCTIONS OF VENUS, 1999–2005
(* = transit)

1999 August 20
2001 March 30
2002 October 31
2004 June 8*

TOTAL SOLAR ECLIPSES, 1999–2005

1999 August 11	England, Europe, Turkey, India
2001 June 21	Atlantic, South Africa
2002 June 10	Pacific
2003 November 23	Antarctic
2005 April 8	Pacific, America

Index

"Filling a very important niche in astronomical literature."
Irish Astronomical Journal

"They all go together to make a very valuable series worth its weight in gold."
Amateur Astronomy

"...a very accessible and modern source of astronomical information."
Journal of the British Astronomical Association

The Practical Astronomy Series - of which
Patrick Moore is the Series Editor - is for anyone who is seriously
interested in the subject, but who may not be scientists, engineers or
astronauts themselves!

All books in the series are written specifically for enthusiasts who have:
- gone beyond the first stages of learning about astronomy,
- who quite probably own, are thinking of buying, or have access to a
  moderately good optical telescope of some kind,
- and who want to read more.

The last few pages of this book are devoted to giving brief information
about one or two of the other titles in this series.

All books are, of course, available from all good booksellers (who can
order them even if they are not in stock), but if you have difficulties you
can contact the publishers direct, by telephoning +44 (0) 1483 418822
(in the UK and Europe), +1/212/4 60 15 00 (in the USA),
or by emailing orders@svl.co.uk

**www.springer.co.uk   www.springer.de
www.springer-ny.com**

Patrick Moore's Millennium Yearbook

Patrick Moore and Allan Chapman

In a year that will probably be remembered almost as much for books about the millennium as for the turn of 2000 AD itself, **Patrick Moore's Millennium Yearbook** celebrates... well, the wrong millennium!

This thoroughly entertaining book - which is for everyone, not not just astronomers - contains articles on King Alfred's chronological work, reviews of the new Star Catalogue by the Arab Al-Sufi and the latest edition of Ptolemy's Almagast. And foreshadowing the change to metric units by 1000 years, the book uses arabic numbers instead of Roman - but there is a conversion table if you have trouble with the idea of "zero" and prefer the older system.

96 pages
Softcover
ISBN 1-85233-619-6

Please see page 165 for ordering details.

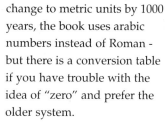

Observing Meteors, Comets, Supernovae and other Transient Phenomena

Neil Bone

Did you ever think of the night sky as unchanging and predictable?

It isn't, of course. Neil Bone has written entirely about observing transient phenomena - events, often short-lived, that can take even the experienced observer by surprise.

In a detailed but thoroughly readable book, he discusses the best way to observe, photograph, or otherwise record transient events associated with: meteors, aurorae and other atmospheric phenomena, the Sun, the Moon, the planets, asteroids and planetary satellites, comets, variable stars, novae and supernovae.

200 pages
Softcover
ISBN 1-85233-017-1

Please see page 165 for ordering details.

Amateur Telescope Making

Edited by Stephen F. Tonkin

If you're interested in ATM, then this is the book for you!

Some of the world's leading experts describe a whole range of construction projects, from "shoe-string" telescopes - in which low cost is the most vital ingredient - to ambitious instruments like the quad-Shiefspiegler.

This book isn't only about telescopes, it's also about mountings and accessories, drives and all sorts of other things you can make for yourself.

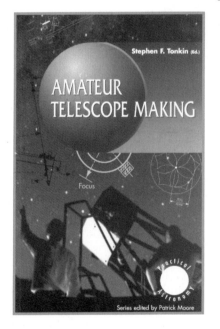

228 pages
Softcover
ISBN 1-85233-000-7

Please see page 165 for
ordering details.